SELECTED TOPICS IN
SHOCK WAVE PHYSICS AND EQUATION OF STATE MODELING

SELECTED TOPICS IN

SHOCK WAVE PHYSICS AND EQUATION OF STATE MODELING

G. Roger Gathers, Ph.D.
Lawrence Livermore National Laboratory
Livermore, CA 94550, USA

Published by

World Scientific Publishing Co. Pte. Ltd.
P O Box 128, Farrer Road, Singapore 9128
USA office: Suite 1B, 1060 Main Street, River Edge, NJ 07661
UK office: 73 Lynton Mead, Totteridge, London N20 8DH

SELECTED TOPICS IN SHOCK WAVE PHYSICS AND EQUATION OF STATE MODELING

ISBN 981-02-1691-2

Printed in Singapore by JBW Printers & Binders Pte. Ltd.

FOREWORD

For many years, the shock wave community has remained sufficiently small that there has been little commercial market for books on shock wave physics. The few books that existed were either very comprehensive and thus quite condensed, written from a mathematicians point of view, or specialized to areas such as chemistry in shock tubes. Books are beginning to appear that treat the uses of shock waves and the processes that occur behind the shock front in solids. The present material is based on notes accumulated from an ongoing seminar on shock wave physics, primarily in solids, taught by the author. Much of the material was also presented in a week long short course at Alabama A & M University in Huntsville, February 8-12, 1988. Various individuals who have obtained copies of the course notes found them a considerable help in getting started in the field, and suggested that they should be put out in a more generally available form. It is assumed that the reader has a background in calculus, thermodynamics, mechanics, and statistical mechanics. The discussion of the soft-sphere equation of state assumes some background in solid state physics (the reader is assumed to be familiar with the Debye model of the solid, the concept of phonons, and the Madelung constant for an atomic lattice.)

Since the material is intended to be introductory, rigor is reduced to that necessary to communicate the basics, and the level of detail provided is in the spirit of the Schaum's Outline Series. In a few cases, such as the development of the Mie-Grüneisen equation of state, rigor is not sacrificed. The range of topics is by no means exhaustive, and is focussed on the interests of an experimentalist rather than a theoretician. Approximate methods are thus presented and measurement techniques are given a good deal of space. Hugoniot measurements have been used to provide data for constructing equations of state for hydrocode simulations, and thus a good bit of material is also included on equation of state modeling. Since a hydrocode must access the equation of state for a material for every cell and every time step in a problem, it is imperative that analytic models be used to describe the equation of state as much as possible in order to reduce the expense of computation.

Some comments about the equation of state models is in order. The selection is intended to show the variety in form and level of sophistication. The Mie-Grüneisen equation of state is a classic model which assumes the atoms in a crystal may be treated as a collection of harmonic oscillators. It is thus given extensive treatment. It is primarily useful for describing solids.

Some of the models are specialized, being intended only for impact of solids on one another (one includes the possibility of energy input via x-ray deposition.) The soft-sphere model is applicable to liquid metals. The treatment of the electronic contributions to the pressure is somewhat ad hoc but it is nonetheless useful for describing experimental data. Perhaps the most elaborate model is the multi-branched analytic equation of state which attempts to accurately model the liquid-vapor two-phase region while at the same time giving good representation for greatly expanded and heated vapor. It was developed for the laser-fusion power program. A recommended source of Hugoniot data for those interested is the compendium published by the shock wave group at Los Alamos National Laboratory[1].

Nomenclature for the shock and mass velocities has two common forms in the literature. The first uses U_s for the shock velocity, and U_p for the particle, or mass velocity. While this is quite specific, it is also somewhat cumbersome to carry the subscripts around in the equations. The author thus prefers the form D for the shock velocity, and u for the mass velocity, a form found in much of the Russian work. The locus of possible states resulting from shock compression is often called the "shock adiabat". This is a very unfortunate choice, since the term adiabatic is so closely associated with isentropic processes, and shock compression is not isentropic. I will attempt to studiously avoid this terminology.

A good bit of the material is original, but much was also gleaned from various review papers, books, and discussions with other workers. It has been accumulated over so long a time that the author is unable to specifically credit all the sources, and thus no attempt will be made. It is hoped that anyone who recognizes a development that was original with them will forgive the lack of acknowledgement. For those with more interest in theory, an excellent review article has been written by Marvin Ross[2]. The topics include Monte Carlo and molecular dynamics calculations and fluid variational theory. Considerable material is also presented on theoretical interpretation of shock compression data of simple liquids (argon, xenon, molecular hydrogen, dissociation of dense liquid nitrogen, carbon compounds, and the structure of dense alkali halide melts.)

The author wishes to specifically thank David A. Young, who read the draft and made numerous helpful suggestions. Any remaining errors are of course, the responsibility of the author.

G. Roger Gathers

TABLE OF CONTENTS

1. Fundamental Results for Plane Shock Waves

1.1 Hugoniot Relations in Lab and Shock Reference Frames.

The Hugoniot relations are a statement of the conservation laws for mass, momentum, and energy. They may be written in various reference frames, but the two most useful are the lab frame and one in which the shock is stationary. A variety of variables may be used for the representation (eg, for work in gases, one sometimes sees them using the Mach number as a variable.) For most work involving solids, however, one commonly sees mass and shock velocities, pressure, specific internal energy, and relative volume as the choice of variables. For use in hydrocode representations, one sees dimensionless variables, such as compression, and specific energy and density referred to standard values. Specification of any two variables determines the rest.

It is important to fix the fact firmly in mind that the equations do not describe a property of any specific material. They are most accurately called the "jump" equations, relating the change in variables across a shock front. The final pressure and relative volume reached depend on the initial conditions present when the shock arrives.

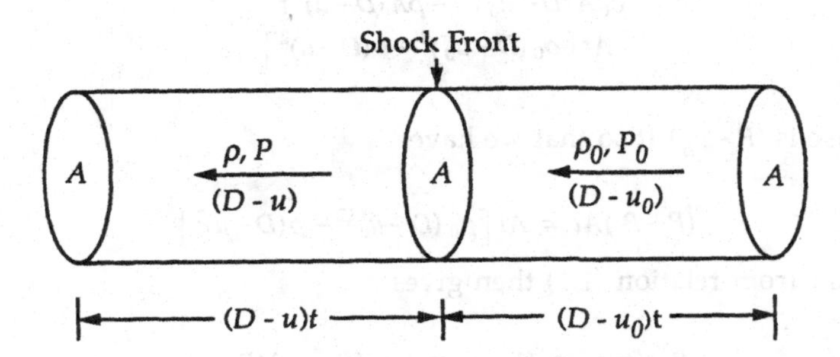

Fig. 1.1. Reference frame with shock at rest. D, u are shock and mass velocities, resp. in the lab frame. Velocities in parentheses are the mass velocities through the shock front. ρ = density, P = pressure, t = time, A = cross section. Subscript zero indicates values in the undisturbed medium.

One can thus have successive shocks driving the material to higher and higher pressures and compression. The form of the equations is the same for each successive shock, but the initial conditions for each case are the final conditions left by the previous shock. Without further preamble, we will begin with the lab frame case.

Figure 1.1 shows the shock surface with a cylinder of mass flowing into and out of it in a given time interval t. The initial density ρ_0, pressure P_0, and mass velocity u_0 are considered as specified parameters. D is the shock velocity, and u is the mass velocity behind the shock, both in lab frame. A is the cross section of the cylinder. The left hand cylinder has length $(D-u)t$ and thus mass $m_l = \rho A(D-u)t$. The length of the right hand cylinder is $(D-u_o)t$ and its mass is $m_r = \rho_0 A(D-u_0)t$.
Conservation of mass thus gives

$$\boxed{\rho(D-u) = \rho_0(D-u_0)} \tag{1.1}$$

The net force across the shock is $(P - P_0)A$, directed toward the right. The momentum change from left to right is

$$\begin{aligned} m_r(D-u_0) - m_l(D-u) &= \\ \rho_0 A(D-u_0)^2 t - \rho A(D-u)^2 t &= \\ At\left[\rho_0(D-u_0)^2 - \rho(D-u)^2\right] \end{aligned}$$

The impulse is $(P - P_0)At$ so that we have

$$(P-P_0)At = At\left[\rho_0(D-u_0)^2 - \rho(D-u)^2\right]$$

Substitution from relation (1.1) then gives

$$P - P_0 = \rho_0(D-u_0)^2 - \rho_0(D-u_0)(D-u)$$

so that we have

$$\boxed{P - P_0 = \rho_0(D-u_0)(u-u_0)} \tag{1.2}$$

From relation (1.1),

$$\rho(D-u) = \rho_0(D-u_0)+\rho u_0-\rho u_0$$

or

$$\rho(u-u_0) = \rho D+\rho_0(u_0-D)-\rho u_0 = (\rho-\rho_0)(D-u_0)$$

We thus have

$$\rho(u-u_0) = \rho_0(D-u_0)\left(\frac{\rho-\rho_0}{\rho_0}\right)$$

and

$$(u-u_0) = \rho_0(D-u_0)\left(\frac{\rho-\rho_0}{\rho\rho_0}\right)$$

so that,

$$\boxed{P-P_0 = \rho_0^2(D-u_0)^2(V_0-V)} \qquad (1.3)$$

It is useful to solve the equations for the mass and shock velocities. We have shown that

$$(u-u_0) = \rho_0(D-u_0)(V_0-V)$$

and from relation (1.3) we have

$$\rho_0^2(D-u_0)^2 = (P-P_0)/(V_0-V)$$

so that

$$(u-u_0)^2 = (P-P_0)(V_0-V)^2/(V_0-V)$$

or

$$(u-u_0)^2 = (P-P_0)(V_0-V)$$

and thus

$$\boxed{u = u_0 \pm \sqrt{(P-P_0)(V_0-V)}} \qquad (1.4)$$

We can also write

$$(D-u_0)^2 = (P-P_0)/\rho_0^2 (V_0 - V)$$

or

$$(D-u_0)^2 = V_0^2 (P-P_0)/(V_0 - V)$$

so that

$$\boxed{D = u_0 \pm V_0 \sqrt{\frac{(P-P_0)}{(V_0 - V)}}} \tag{1.5}$$

The negative radicals correspond to backward going waves. The increase in the total energy of the medium is equal to the net work done on it. On the left we have $m_l = \rho A(D-u)t$ with total specific energy $E + u^2/2$ and on the right we have $m_r = \rho_0 A(D-u_0)t$ with total specific energy $E_0 + u_0^2/2$. The total increase in energy is thus

$$\rho A(D-u)t\left(E+\frac{u^2}{2}\right) - \rho_0 A(D-u_0)t\left(E_0 + \frac{u_0^2}{2}\right)$$

The net work done is the work done on the medium at the face on the right minus the work done by the medium at the face on the left:

$$W = (Pu - P_0 u_0)At$$

We thus have

$$\rho(D-u)\left(E+\frac{u^2}{2}\right) - \rho_0 (D-u_0)\left(E_0 + \frac{u_0^2}{2}\right) = Pu - P_0 u_0$$

Conservation of mass however, allows us to write

$$\rho_0 (D-u_0)\left[E - E_0 + \frac{u^2}{2} - \frac{u_0^2}{2}\right] = Pu - P_0 u_0$$

so that

$$E - E_0 = \frac{Pu - P_0 u_0}{\rho_0 (D-u_0)} - \frac{1}{2}(u^2 - u_0^2)$$

Consider a forward-going wave for convenience. From relation (1.5) we can write

$$\rho_0 (D-u_0) = (P-P_0)^{1/2} (V_0 - V)^{-1/2}$$

and from relation (1.4)

$$u - u_0 = (P - P_0)^{1/2}(V_0 - V)^{1/2}$$

Manipulation of this gives

$$Pu - P_0 u_0 = u_0(P - P_0) + P(P - P_0)^{1/2}(V_0 - V)^{1/2}$$

Also, we have

$$\frac{1}{2}\left(u^2 - u_0^2\right) = u_0(P - P_0)^{1/2}(V_0 - V)^{1/2} + (P - P_0)(V_0 - V)$$

Substitution in the energy equation leads to

$$\boxed{E - E_0 = \frac{1}{2}(P + P_0)(V_0 - V)} \qquad (1.6)$$

Now consider the frame where the shock wave is stationary. The conservation relations take on a form that indicates the flux of the conserved quantities through the shock front is constant. First consider the relations (1.1, 1.2, and 1.6) for the lab frame. Let v be the velocity of the material exiting the front, and v_0 the velocity of the material entering the front. We have

$$v = D - u$$

and

$$v_0 = D - u_0$$

Substituting in relation (1.1) we have

$$\boxed{\rho v = \rho_0 v_0} \qquad (1.7)$$

Now

$$v_0 - v = D - u_0 - D + u = u - u_0$$

so that we may substitute in relation (1.2) to give

$$P - P_0 = \rho_0 v_0 (v_0 - v)$$

But since $\rho v = \rho_0 v_0$, we can also write

$$P - P_0 = \rho_0 v_0^2 - \rho v^2$$

or

$$\boxed{P + \rho v^2 = P_0 + \rho_0 v_0^2} \tag{1.8}$$

Now consider the energy equation. Solve for $P + P_0$

$$\begin{aligned} P + P_0 &= 2(E - E_0)/(V_0 - V) \\ &= 2(E - E_0)/\left(\frac{1}{\rho_0} - \frac{1}{\rho}\right) \\ &= 2\rho_0 (E - E_0)/\left(1 - \frac{\rho_0}{\rho}\right) \end{aligned}$$

or using relation (1.7)

$$P + P_0 = 2\rho_0 (E_0 - E)/(v/v_0 - 1) \tag{1.9}$$

But from relation (1.8)

$$P - P_0 = \rho_0 v_0^2 - \rho v^2 \tag{1.10}$$

Add and subtract relations (1.9) and (1.10) to solve for P and P_0.

$$2P = 2\rho_0 (E_0 - E)/(v/v_0 - 1) + \rho_0 v_0^2 - \rho v^2 \tag{1.11}$$

$$2P_0 = 2\rho_0 (E_0 - E)/(v/v_0 - 1) - \rho_0 v_0^2 + \rho v^2 \tag{1.12}$$

Divide relation (1.11) by 2ρ and relation (1.12) by $2\rho_0$ and subtract the resulting equations:

$$\begin{aligned} P/\rho - P_0/\rho_0 &= (\rho_0/\rho)(E_0 - E)/(v/v_0 - 1) - (E_0 - E)/(v/v_0 - 1) \\ &\quad + \rho_0 v_0^2/2\rho - v^2/2 + v_0^2/2 - \rho v^2/2\rho_0 \end{aligned}$$

or

$$P/\rho - P_0/\rho_0 = (\rho_0/\rho - 1)(E_0 - E)/(v/v_0 - 1) + v_0^2/2$$
$$-v^2/2 + \left(\rho_0 v_0^2/\rho - \rho v^2/\rho_0\right)/2$$

From relation (1.7) the first term on the right is just $E_0 - E$.

Consider the last term:

$$\left(\rho_0 v_0^2/\rho - \rho v^2/\rho_0\right) = (v v_0 - v_0 v) = 0$$

so that

$$P/\rho - P_0/\rho_0 = E_0 - E + v_0^2/2 - v^2/2$$

Finally, we have

$$\boxed{E + \frac{P}{\rho} + \frac{v^2}{2} = E_0 + \frac{P_0}{\rho_0} + \frac{v_0^2}{2}} \tag{1.13}$$

If we write specific enthalpy $h = E + PV = E + P/\rho$ the energy conservation relation simplifies even further to

$$\boxed{h + \frac{v^2}{2} = h_0 + \frac{v_0^2}{2}} \tag{1.14}$$

$$\rho v = \text{mass flux}$$
$$P + \rho v^2 = \text{momentum flux}$$
$$E + P/\rho + v^2/2 = \text{total specific energy}$$

1.2 Shock Velocity - Particle Velocity Relations

The locus of possible final states due to shock compression for a material corresponding to initial conditions of normal density, and zero mass velocity and pressure is called the principal Hugoniot, and is determined by the equation of state of the material. It can be calculated from the equation of state by applying the constraints given by the Hugoniot equations. In general, the principal Hugoniot must be experimentally measured. Numerous experimental techniques have been developed for this purpose. Theoretical models such as the Grüneisen equation of state can then be used to interpolate from the measured states. In all further discussions, reference to the Hugoniot will mean the principal Hugoniot unless specified otherwise (eg; reshock Hugoniots.) There is only one principal Hugoniot for a material, but an infinite number of reshock Hugoniots, since the latter can begin at any point of the principal Hugoniot. For most materials, a simple linear relationship between the particle velocity u and the shock velocity D is observed approximately over a considerable pressure range:

$$D = c_0 + su \tag{1.15}$$

The intercept c_0 does not always correspond to the normal sound speed. Where phase changes are involved, the Hugoniot may be represented by a series of such straight line segments, with each segment corresponding to a particular range of u. For simple materials, we will see that the slope s is determined by the derivative of the bulk modulus at zero pressure.

In some cases the departure from linearity is enough to justify the use of a quadratic, and occasionally a cubic fit. The linear case is especially useful, since it results in a quadratic dependence of pressure on particle velocity. This makes analytical solution of collision problems easy, since the form of the solution of a quadratic is well known and simple. If we substitute relation (1.15) in the momentum conservation relation with zero initial mass velocity, we obtain

$$P = \rho_0 (c_0 + su)u \tag{1.16}$$

Many attempts have been made to develop a theoretical explanation of why the linear relation between shock and mass velocities is so commonly observed, but all such attempts seem to have the flavor of the efforts to determine why the fine structure constant is so close to 1/137. Walker,

Walker, and Walker[3] have devised a scheme for predicting approximate Hugoniots for a wide range of elements using initial density ρ_0, atomic radius r_a, and atomic weight ω_a. They cite correlations drawn from comparisons of data and Hugoniots calculated with molecular dynamics methods. Their basic form is

$$D = f(P)(r_a/\omega_a)^{1/4}\rho_0^{-0.1}$$

where $f(P)$ is an adjusting function, D is in mm/ms, r_a is in Angstroms, ω_a is in g/mole, density is in g/cm^3, and P is in megabars. The general $f(P)$ for all elements considered (Be, Ti, Fe, Ni, Ag, Ta, W, Pt, Au, U, Mg, Al, Cu, Zn, Li, Na, K, Bi, and Pb) is

$$f(P) = 0.42P + 10.3P^{1/2} + 12$$

For $P \leq 1.6$ MBar, for Mg, Al, Cu, and Zn, an improved fit results for

$$f(P) = 12.4P - 0.37P^{1/2} + 12$$

while for elements with large r_a (Na, K, Pb, Bi) better fits to sonic velocities are obtained with

$$f(P) = 3.7 + 12.1P^{1/2} + 6$$

It should be emphasized that these fits as such, give no special insight into the real causes of the form. They merely describe observed correlations.

Walzer[4] has used quantum mechanics to make approximate calculations for the transition metals to derive the principal Hugoniot, bulk modulus, pressure derivatives, etc. Nine input parameters are used (atomic radius, core radius, d - state radius, nearest neighbor separation, isothermal bulk modulus at zero pressure, etc.) The d - band structure has been expressed through a combination of the transition metal pseudopotential theory and muffin-tin orbital theory. The theory is applicable in a unified manner to all transition metals. Only two of the input parameters are macrophysical (Bulk modulus and Grüneisen parameter at zero pressure.)

The principal Hugoniot for the case of a linear shock velocity - mass velocity relationship can be written in a much-used simple analytic form.

Define the dimensionless parameter ξ as the ratio of the mass and shock velocities:

$$\xi = u/D = 1 - \rho_0/\rho$$

We can then write

$$D = c_0 + su = c_0 + s(D\xi)$$

$$D(1 - s\xi) = c_0$$

or

$$D = \frac{c_0}{(1 - s\xi)} \tag{1.17}$$

Now we have

$$u = (D - c_0)/s = c_0 s\xi / s(1 - s\xi)$$

so that

$$u = \frac{c_0 \xi}{(1 - s\xi)} \tag{1.18}$$

Substitution in the momentum relation for lab frame with u_0 and P_0 equal to zero gives

$$P = \frac{\rho_0 c_0^2 \xi}{(1 - s\xi)^2} \tag{1.19}$$

This gives the principal Hugoniot in terms of the compression parameter ξ. Direct differentiation with respect to ξ then gives

$$\frac{dP}{d\xi} = \rho_0 c_0^2 \frac{(1 + s\xi)}{(1 - s\xi)^3} \qquad 1.(20)$$

Now

$$\frac{dP}{d\rho} = \left(\frac{d\xi}{d\rho}\right)\left(\frac{dP}{d\xi}\right)$$

and

$$\frac{d\xi}{d\rho} = \frac{(1-\xi)^2}{\rho_0}, \quad \frac{dP}{d\rho} = \frac{(1-\xi)^2}{\rho_0}\frac{dP}{d\xi}$$

so that

$$\left(\frac{dP}{d\rho}\right)_H = c_0^2\frac{(1+s\xi)(1-\xi)^2}{(1-s\xi)^3} \tag{1.21}$$

Now the bulk modulus on the Hugoniot B_H is given by

$$B_H = \rho\left(\frac{\partial P}{\partial \rho}\right)_H$$

and

$$\rho = \frac{\rho_0}{(1-\xi)}$$

so that

$$\rho\left(\frac{\partial P}{\partial \rho}\right)_H = \frac{\rho_0 c_0^2(1+s\xi)(1-\xi)}{(1-s\xi)^3}$$

and thus

$$B_H = \rho_0 c_0^2\frac{(1-\xi)(1+s\xi)}{(1-s\xi)^3} \tag{1.22}$$

1.3 Relation of Shock velocity-Particle Velocity Slope to the Derivative of the Bulk Modulus at Zero Pressure

Now for initial conditions $\rho/\rho_0 = 1$ and thus $\xi = 0$, and therefore

$$B_{0H} = \rho_0 c_0^2$$
$$c_0 = \sqrt{B_{0H}/\rho_0}$$

But at initial conditions, the Hugoniot and isentrope are tangent to each other and have the same curvature. We can thus write

$$B_{0H} = B_{0S}\,, \quad B_{0H}' = B_{0S}'\,,$$

hence c_0 is just the adiabatic sound speed at initial conditions for this model. Now

$$B_{0H}' = (dB_H/dP) \quad \text{at zero pressure, or}$$

$$B_{0H}' = (d\rho/dP)(d\xi/d\rho)(dB_H/d\xi) \quad \text{for } P = 0.$$

But we have already shown that

$$\frac{d\xi}{d\rho} = \frac{(1-\xi)^2}{\rho_0}, \quad \frac{\rho}{\rho_0} = \frac{1}{(1-\xi)}$$

and

$$\left(\frac{d\rho}{dP}\right)_H = \frac{\rho}{B_H}$$

so that

$$\frac{dB_H}{dP} = \frac{\rho}{\rho_0}\frac{(1-\xi)^2}{B_H}\left(\frac{dB_H}{d\xi}\right) = \frac{(1-\xi)}{B_H}\left(\frac{dB_H}{d\xi}\right)$$

Now

$$B_H = \rho_0 c_0^2 \frac{(1-\xi)(1+s\xi)}{(1-s\xi)^3}$$

so that

$$\frac{dB_H}{d\xi} = \rho_0 c_0^2 \left[\frac{(4s-1)+s\xi(2s-s\xi-4)}{(1+s\xi)^4}\right]$$

and thus

$$\frac{dB_H}{dP} = \left[\frac{(4s-1)+s\xi(2s-s\xi-4)}{1-s^2\xi^2}\right] \tag{1.23}$$

For $P = 0$ we thus have $\xi = 0$ and since $B_{0H}' = B_{0S}'$ we have

$$B_{0_s}' = 4s - 1$$

and thus

$$\boxed{s = \frac{1}{4}(B_{0_s}' + 1)} \tag{1.24}$$

We thus see that the slope parameter in the shock velocity - mass velocity relationship is determined by the pressure derivative of the adiabatic bulk modulus at zero pressure.

1.4 Energy Partition in a Single Shock Wave

Examination of the momentum conservation relations (1.2), and (1.3) shows that for a given shock strength (hence given shock speed D), the momentum conservation condition connects the initial and final states with a straight line in either the P, V plane, or the P, u plane. The line is called the Rayleigh line. It should be noted that the physical processes in the shock front do not need to follow the Rayleigh line. It only connects the initial and final states.

We may examine energy partition conveniently in the P, V plane. Consider the relation (1.4) for energy conservation, and relation (1.5) for the particle velocity. We will assume the material is initially at rest, so that $u_0 = 0$. We then have

$$\frac{u^2}{2} = \frac{1}{2}(P - P_0)(V_0 - V)$$

This is the kinetic energy/unit mass acquired in the compression and is thus equal to the area of the triangle bcd in figure 1.2.

Now the total specific energy acquired is

$$\frac{u^2}{2} + E - E_0 = \frac{1}{2}(P - P_0)(V_0 - V) + \frac{1}{2}(P + P_0)(V_0 - V) = P(V_0 - V)$$

The total specific energy acquired by the material is thus equal to the area of the rectangle acde. We may thus conclude that the shaded region represents

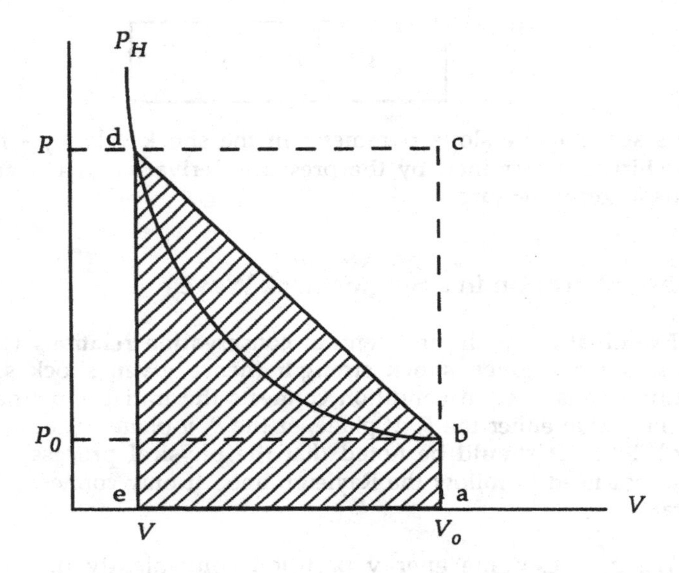

Fig. 1.2. Energy partition in a single shock. b-d is called the Rayleigh line. The shaded region represents the specific internal energy acquired in the compression. Area bcd is the kinetic energy/unit mass.

the specific internal energy acquired in the compression. The Rayleigh line is a possible path for the shock compression. In some situations, such as those where the shock has finite thickness due to rate limited processes, the actual path followed may be significantly different from the Rayleigh line. Notice also, that if the initial pressure P_0 is negligible, equipartition of energy between kinetic and internal energy occurs.

Now consider isentropic compression, rather than shock compression. The Hugoniot always passes above the isentrope from a given state (for materials such that the bulk coefficient of thermal expansion at constant pressure is positive.)

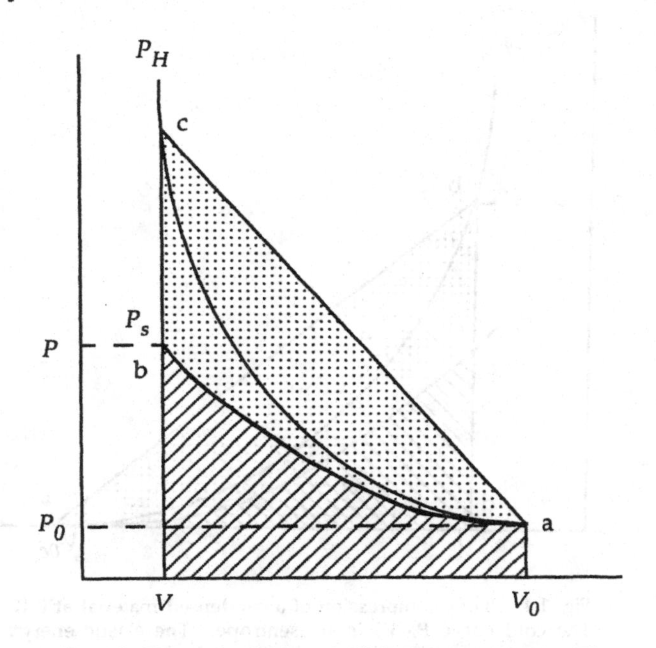

Fig. 1.3. Energy in shock and adiabatic compressions. The work done in the isentropic process is equal to the shaded area. The heat added is equal to the area of the stippled region abc.

Figure 1.3 illustrates the situation. The work done in the isentropic process is equal to the cross-hatched area. As before, this also gives the increase in internal energy of the material. In order to reach the final state at c, it is necessary to heat the material. The heat added is equal to the stippled area abc. This area also determines the entropy increase of the material due to shock compression.

Now consider shock compression of a cold (0 K) condensed material. Figure 1.4 illustrates the process. The cold curve $P_c(V)$ is an isentrope. The elastic energy E_c acquired by the material is equal to the cross-hatched area.

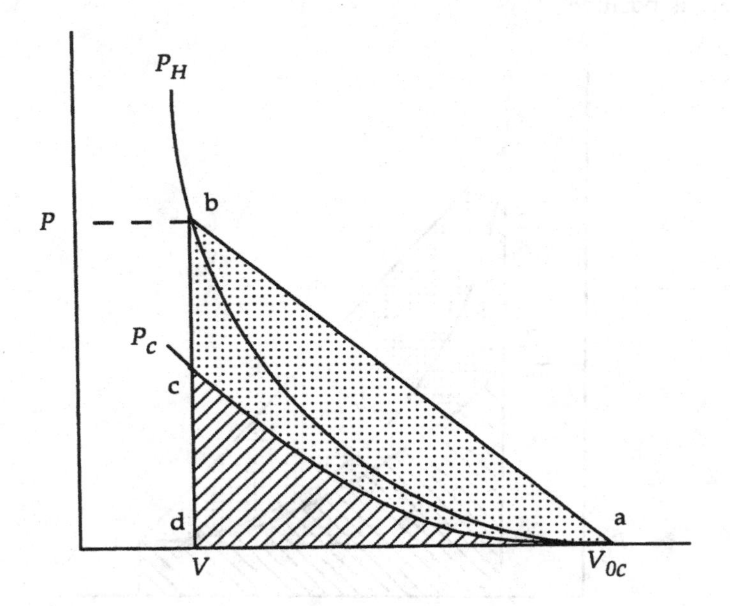

Fig. 1.4. Shock compression of a condensed material at 0 K. The cold curve $P_c(V)$ is an isentrope. The elastic energy acquired by the material is equal to the shaded area. The total internal energy is given by the triangle abd. The stippled area gives the thermal part of the internal energy.

$$E_c(V) = \int_V^{V_{0c}} P_c(V)dV$$

The total internal energy E is (for P_0 negligible in relation (1.4))

$$E = \frac{1}{2}P(V_{0c} - V)$$

which is given by the the area of the triangle abd. The stippled area thus represents the thermal part of the internal energy.

Now consider shock compression of a solid from standard conditions. Figure 1.5 illustrates the situation.

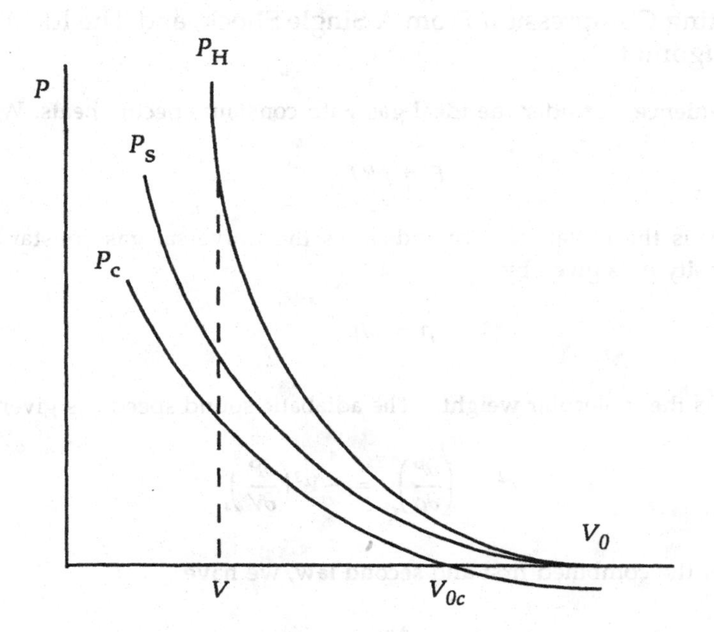

Fig. 1.5. Compression of a solid from standard conditions. P_H is the Hugoniot and P_s is the isentrope, both emanating from standard conditions. P_c is the cold curve, emanating from 0 K.

The Hugoniot and isentrope are shown emanating from standard conditions. The cold curve is shown below them. The initial elastic pressure is negative at V_0 . The thermal pressure involved in raising the temperature from 0 degrees K to T_0 (and hence from V_{0c} to V_0) causes the mean atomic separation to be greater than it would be at 0 K, resulting in tension from the interatomic forces.

For very strong shocks (megabar range) the difference between V_0 and V_{0c} , and between the isentrope and cold compression curve become unimportant, while the difference between the Hugoniot and these curves becomes very appreciable. The thermal pressure becomes comparable to the total pressure.

1.5 Limiting Compression From A Single Shock, and The Ideal Gas Hugoniot

For convenience, consider the ideal gas with constant specific heats. We have

$$P = \bar{\rho} RT$$

where $\bar{\rho}$ is the molar density and R is the universal gas constant. The mass density ρ is given by

$$\rho = M\bar{\rho}$$

where M is the molecular weight. The adiabatic sound speed c is given by

$$c^2 = \left(\frac{\partial P}{\partial \rho}\right)_s = -V^2\left(\frac{\partial P}{\partial V}\right)_s$$

Now from the combined first and second law, we have

$$TdS = \frac{\kappa_T c_v}{\beta} dP + \frac{c_p}{\beta V} dV$$

where

$$\beta = \frac{1}{V}\left(\frac{\partial P}{\partial T}\right)_p, \quad \kappa_T = -\frac{1}{V}\left(\frac{\partial V}{\partial P}\right)_T$$

For an adiabatic process $dS = 0$ so that

$$\kappa_T c_v dP_s = -\frac{c_p}{V} dV_s$$

$$\left(\frac{\partial P}{\partial V}\right)_s = -\frac{c_p}{\kappa_T c_v}\frac{1}{V} = -\frac{\gamma}{\kappa_T V}$$

$$-V^2\left(\frac{\partial P}{\partial V}\right)_s = \frac{\gamma V}{\kappa_T} = c^2$$

But

$$\kappa_T = -\frac{1}{V}\left(\frac{\partial V}{\partial P}\right)_T$$

so that

$$c^2 = -\gamma V^2\left(\frac{\partial P}{\partial V}\right)_T$$

But from the equation of state

$$P = \left(\frac{R}{M}\right)\frac{T}{V}$$

so that

$$\left(\frac{\partial P}{\partial V}\right)_T = -\left(\frac{R}{M}\right)\frac{T}{V^2}$$

and

$$-V^2\left(\frac{\partial P}{\partial V}\right)_T = \frac{R}{M}T$$

so that we can write

$$\boxed{c = \sqrt{\frac{\gamma RT}{M}}} \qquad (1.25)$$

We can also write

$$c^2 = c_o^2\left(\frac{T}{T_o}\right) \tag{1.26}$$

But since $PV = RT/M$, we can also write

$$c^2 = \gamma PV \tag{1.27}$$

Now consider the Hugoniot equations in lab frame.

$$\begin{aligned}
\rho_0 D &= \rho(D-u) \\
P-P_0 &= \rho_0 Du \\
E-E_0 &= \frac{1}{2}(P+P_0)\left(\frac{1}{\rho_0}-\frac{1}{\rho}\right)
\end{aligned}$$

If we consider an adiabatic process for an ideal gas, we can show that

$$\begin{aligned}
E &= \frac{PV}{(\gamma-1)} = \frac{P}{(\gamma-1)\rho} \\
E-E_0 &= \frac{1}{(\gamma-1)}\left(\frac{P}{\rho}-\frac{P_0}{\rho_0}\right)
\end{aligned}$$

thus

$$\frac{1}{(\gamma-1)}\left(\frac{P}{\rho}-\frac{P_0}{\rho_0}\right) = \frac{1}{2}(P+P_0)\left(\frac{1}{\rho_0}-\frac{1}{\rho}\right)$$

Define $\mathbf{S} \equiv P/P_0$ (shock "strength" parameter)

$$\begin{aligned}
\frac{1}{(\gamma-1)}P_0\left(\frac{\mathbf{S}}{\rho}-\frac{1}{\rho_0}\right) &= \frac{1}{2}P_0(\mathbf{S}+1)\left(\frac{1}{\rho_0}-\frac{1}{\rho}\right) \\
\frac{1}{(\gamma-1)}\frac{\mathbf{S}}{\rho} - \frac{1}{(\gamma-1)}\frac{1}{\rho_0} &= \frac{(\mathbf{S}+1)}{2\rho_0} - \frac{(\mathbf{S}+1)}{2\rho}
\end{aligned}$$

$$\left[\frac{S}{(\gamma-1)}+\frac{(S+1)}{2}\right]\frac{1}{\rho} = \left[\frac{(S+1)}{2}+\frac{1}{(\gamma-1)}\right]\frac{1}{\rho_0}$$

$$\frac{\rho}{\rho_0} = \frac{\left[S/(\gamma-1)+(S+1)/2\right]}{\left[(S+1)/2+1/(\gamma-1)\right]} = \frac{2S+(S+1)(\gamma-1)}{(S+1)(\gamma-1)+2}$$

Or finally,

$$\boxed{\frac{\rho}{\rho_0} = \frac{S(\gamma+1)+(\gamma-1)}{S(\gamma-1)+(\gamma+1)}} \tag{1.28}$$

Now conservation of mass gives

$$\frac{\rho_0}{\rho} = \frac{D-u}{D} = 1-\frac{u}{D} \quad \text{so that} \quad \frac{u}{D} = 1-\frac{\rho_0}{\rho}$$

$$\frac{u}{D} = \frac{2(S-1)}{S(\gamma+1)+(\gamma-1)}$$

or

$$D = u\frac{\left[S(\gamma+1)+(\gamma-1)\right]}{2(S-1)}$$

But $P - P_0 = \rho D u$ so that

$$P-P_0 = \rho u^2\frac{\left[S(\gamma+1)+(\gamma-1)\right]}{2(S-1)}$$

Divide by P_0:

$$\frac{P}{P_0}-1 = (S-1) = \frac{\rho_0 u^2}{P_0}\frac{\left[S(\gamma+1)+(\gamma-1)\right]}{2(S-1)}$$

so that

$$u^2 = \frac{2(S-1)^2}{\left[S(\gamma+1)+(\gamma-1)\right]}\frac{P_0}{\rho_0}$$

and thus

$$\boxed{u = (\mathbf{S}-1)\sqrt{\frac{2P_0V_0}{\mathbf{S}(\gamma+1)+(\gamma-1)}}} \tag{1.29}$$

$$D = \frac{[\mathbf{S}(\gamma+1)+(\gamma-1)](\mathbf{S}-1)}{2(\mathbf{S}-1)}\sqrt{\frac{2P_0V_0}{[\mathbf{S}(\gamma+1)+(\gamma-1)]}}$$

or

$$\boxed{D = \sqrt{\frac{[\mathbf{S}(\gamma+1)+(\gamma-1)]}{2}P_0V_0}} \tag{1.30}$$

Now $c^2 = \gamma PV$ but we want to calculate this on the Hugoniot, to correspond to the state behind the shock.

$$\frac{P}{P_0}\frac{V}{V_0} = \frac{c^2/\gamma}{P_0V_0} = \mathbf{S}\left[\frac{\mathbf{S}(\gamma-1)+(\gamma+1)}{\mathbf{S}(\gamma+1)+(\gamma-1)}\right]$$

so that

$$c = \sqrt{\frac{\mathbf{S}\gamma[\mathbf{S}(\gamma-1)+(\gamma+1)]}{[\mathbf{S}(\gamma+1)+(\gamma-1)]}P_0V_0}$$

Now since $c_0{}^2 = \gamma P_0V_0$ we can write

$$u = (\mathbf{S}-1)\sqrt{\frac{2}{\gamma[\mathbf{S}(\gamma+1)+(\gamma-1)]}}c_0 \tag{1.31}$$

$$D = \sqrt{\frac{\mathbf{S}(\gamma+1)+(\gamma-1)}{2\gamma}}c_0 \tag{1.32}$$

$$c = \sqrt{\frac{\mathbf{S}[\mathbf{S}(\gamma-1)+(\gamma+1)]}{\mathbf{S}(\gamma+1)+(\gamma-1)}}c_0 \tag{1.33}$$

It is also of interest to calculate

$$b^2 = \left(\frac{\partial P}{\partial \rho}\right)_H = -V^2\left(\frac{\partial P}{\partial V}\right)_H$$

where the subscript indicates the derivative on the Hugoniot. We have

$$E = \frac{PV}{(\gamma-1)}$$

and

$$E - E_0 = \frac{1}{2}(P+P_0)(V_0 - V)$$

$$\frac{PV - P_0 V_0}{(\gamma-1)} = \frac{1}{2}(P+P_0)(V_0 - V)$$

$$PV - P_0 V_0 = \frac{(\gamma-1)}{2}(PV_0 + P_0 V_0 - PV - P_0 V)$$

$$2PV - 2P_0 V_0 = (\gamma-1)PV_0 + (\gamma-1)P_0 V_0 - (\gamma-1)PV - (\gamma-1)P_0 V$$

$$P\left[2V - (\gamma-1)V_0 + (\gamma-1)V\right] = P_0\left[2V_0 + (\gamma-1)V_0 - (\gamma-1)V\right]$$

$$P\left[(\gamma+1)V - (\gamma-1)V_0\right] = P_0\left[(\gamma+1)V_0 - (\gamma-1)V\right]$$

or

$$\boxed{P = \frac{\left[(\gamma+1)V_0 - (\gamma-1)V\right]}{\left[(\gamma+1)V - (\gamma-1)V_0\right]}P_0} \qquad (1.34)$$

This is the Hugoniot for the ideal gas in the P, V plane.

Now

$$\left(\frac{\partial P}{\partial V}\right)_H = \frac{-(\gamma-1)P_0}{\left[(\gamma+1)V - (\gamma-1)V_0\right]} - \frac{\left[(\gamma+1)V_0 - (\gamma-1)V\right]P_0(\gamma+1)}{\left[(\gamma+1)V - (\gamma-1)V_0\right]^2}$$

which reduces to

$$\left(\frac{\partial P}{\partial V}\right)_H = \frac{-4\gamma P_0 V_0}{\left[(\gamma+1)V-(\gamma-1)V_0\right]^2}$$

or, since $c_0{}^2 = \gamma P_0 V_0$,

$$\left(\frac{\partial P}{\partial V}\right)_H = \frac{-4c_0^2}{\left[(\gamma+1)V-(\gamma-1)V_0\right]^2}$$

We may similarly calculate the second derivative:

$$\left(\frac{\partial^2 P}{\partial V^2}\right)_H = \frac{8(\gamma+1)c_0^2}{\left[(\gamma+1)V-(\gamma-1)V_0\right]^3}$$

Now examination of relation (1.28) as $S \to \infty$ shows that the limiting compression is

$$\left(\frac{\rho}{\rho_0}\right)_L = \left(\frac{\gamma+1}{\gamma-1}\right) = \frac{V_0}{V_L}$$

so that

$$(\gamma+1)V_L = (\gamma-1)V_0$$

and $V > V_L$ on the Hugoniot, so that

$$(\gamma+1)V-(\gamma-1)V_0 > 0$$

and thus

$$\left(\frac{\partial^2 P}{\partial V^2}\right)_H > 0$$

The ideal gas Hugoniot is thus concave upward in the P, V plane.

Now

$$b^2 = -V^2\left(\frac{\partial P}{\partial V}\right)_H = \frac{4V^2c_0^2}{\left[(\gamma+1)V-(\gamma-1)V_0\right]^2}$$

$$= \frac{4c_0^2}{\left[(\gamma+1)-(\gamma-1)(V_0/V)\right]^2}$$

But from relation (1.28)

$$\frac{V_0}{V} = \frac{\mathbf{S}(\gamma+1)+(\gamma-1)}{\mathbf{S}(\gamma-1)+(\gamma+1)}$$

and thus

$$(\gamma-1)\left(\frac{V_0}{V}\right) = \frac{\mathbf{S}(\gamma^2-1)+(\gamma-1)^2}{\mathbf{S}(\gamma-1)+(\gamma+1)}$$

and

$$\left[(\gamma+1)-(\gamma-1)(V_0/V)\right] =$$

$$\frac{(\gamma+1)\left[\mathbf{S}(\gamma-1)+(\gamma+1)\right]-\mathbf{S}(\gamma^2-1)-(\gamma-1)^2}{\mathbf{S}(\gamma-1)+(\gamma+1)} = \frac{4\gamma}{\mathbf{S}(\gamma-1)+(\gamma+1)}$$

so that

$$b^2 = \frac{c_0^2\left[\mathbf{S}(\gamma-1)+(\gamma+1)\right]^2}{4\gamma^2}$$

and thus

$$\boxed{b = \frac{\mathbf{S}(\gamma-1)+(\gamma+1)}{2\gamma}c_0} \qquad (1.35)$$

Now calculate the form of the ideal gas Hugoniot in the D, u plane. From relation (1.32) we have

$$\left(\frac{D}{c_0}\right)^2 = \frac{\mathbf{S}(\gamma+1)+(\gamma-1)}{2\gamma} \qquad (1.36)$$

and from relation (1.31):

$$\left(\frac{u}{c_0}\right)^2 = \frac{2(\mathbf{S}-1)^2}{\gamma\left[\mathbf{S}(\gamma+1)+(\gamma-1)\right]} \qquad (1.37)$$

Solve relation (1.36) for $\mathbf{S}$ and substitute in relation (1.37) to eliminate $\mathbf{S}$:

$$\mathbf{S} = \frac{2\gamma(D/c_0)^2-(\gamma-1)}{(\gamma+1)}$$

$$(\mathbf{S}-1) = \frac{2\gamma}{(\gamma+1)}\left(\frac{D}{c_0}\right)^2 + \frac{(-\gamma+1-\gamma-1)}{(\gamma+1)} = \frac{2\gamma}{(\gamma+1)}\left(\frac{D}{c_0}\right)^2 - \frac{2\gamma}{(\gamma+1)}$$

$$(\mathbf{S}-1)^2 = \left(\frac{2\gamma}{\gamma+1}\right)^2\left[(D/c_0)^2-1\right]^2$$

and

$$\gamma\left[\mathbf{S}(\gamma+1)+(\gamma-1)\right] = 2\gamma^2(D/c_0)^2$$

then

$$\left(\frac{u}{c_0}\right)^2 = \frac{2\left(\frac{2\gamma}{\gamma+1}\right)^2\left[\left(\frac{D}{c_0}\right)^2-1\right]^2}{2\gamma^2\left(\frac{D}{c_0}\right)^2} = \frac{4\left[\left(\frac{D}{c_0}\right)^2-1\right]^2}{(\gamma+1)^2\left(\frac{D}{c_0}\right)^2}$$

Let $x \equiv \left(\frac{D}{c_0}\right)^2$, $y \equiv \left(\frac{u}{c_0}\right)^2$ then we have

$$y = \frac{4(x-1)^2}{(\gamma+1)^2 x} \quad \text{or} \quad (x-1)^2 = \left(\frac{\gamma+1}{2}\right)^2 xy$$

$$x^2-2x+1 = \left(\frac{\gamma+1}{2}\right)^2 xy$$

$$x^2-\left[2+\left(\frac{\gamma+1}{2}\right)^2 y\right]x+1 = 0$$

Applying the quadratic formula gives

$$x = 1+\left(\frac{\gamma+1}{2}\right)^2\frac{y}{2}\pm\sqrt{\left[1+\left(\frac{\gamma+1}{2}\right)^2\frac{y}{2}\right]^2-1}$$

or

$$D^2 = c_0^2+\frac{(\gamma+1)^2}{8}u^2+\sqrt{\left[c_0^2+\frac{(\gamma+1)^2}{8}u^2\right]^2-c_0^4}$$

(Simple experiment with trial values verifies that the positive radical is the correct choice.)

Now consider

$$D = A+\sqrt{B}$$

$$D^2 = A^2 + 2A\sqrt{B} + B$$

Let

$$A^2 + B = c_0^2 + \frac{(\gamma+1)^2}{8}u^2$$

and

$$2A\sqrt{B} = \sqrt{\left[c_0^2 + \frac{(\gamma+1)^2}{8}u^2\right]^2 - c_0^4}$$

Solve the last two equations for A and B.

We have

$$4A^2B = \left[c_0^2 + \frac{(\gamma+1)^2}{8}u^2\right]^2 - c_0^4$$

and

$$B = c_0^2 + \frac{(\gamma+1)^2}{8}u^2 - A^2$$

Then eliminating B, we have

$$4A^2B = 4A^2c_0^2 + 4A^2\frac{(\gamma+1)^2}{8}u^2 - 4A^4$$

$$\left[c_0^2 + \frac{(\gamma+1)^2}{8}u^2\right]^2 - c_0^4 = 4A^2c_0^2 + 4A^2\frac{(\gamma+1)^2}{8}u^2 - 4A^4 = 4A^2\left[c_0^2 + \frac{(\gamma+1)^2}{8}u^2\right] - 4A^4$$

Now let $z \equiv A^2$ so that we can write

$$z^2 - \left[c_0^2 + \frac{(\gamma+1)^2}{8}u^2\right]z + \frac{1}{4}\left[c_0^2 + \frac{(\gamma+1)^2}{8}u^2\right]^2 - \frac{c_0^4}{4} = 0$$

Solve for z:

$$z = \frac{\left[c_0^2 + \frac{(\gamma+1)^2}{8}u^2\right] \pm \sqrt{\left[c_0^2 + \frac{(\gamma+1)^2}{8}u^2\right]^2 - \left[c_0^2 + \frac{(\gamma+1)^2}{8}u^2\right]^2 + c_0^4}}{2}$$

or

$$z = \frac{c_0^2 + \frac{(\gamma+1)^2}{8}u^2 \pm c_0^2}{2}$$

$$z^+ = c_0^2 + \frac{(\gamma+1)^2}{16}u^2, \qquad z^- = \frac{(\gamma+1)^2}{16}u^2$$

$$A^+ = \sqrt{c_0^2 + \frac{(\gamma+1)^2}{16}u^2}, \quad A^- = \frac{(\gamma+1)}{4}u$$

$$B^+ = c_0^2 + \frac{(\gamma+1)^2}{8}u^2 - A^{+2}$$

so that

$$B^+ = c_0^2 + \frac{(\gamma+1)^2}{8}u^2 - \left[c_0^2 + \frac{(\gamma+1)^2}{16}u^2\right] = \frac{(\gamma+1)^2}{16}u^2$$

also

$$B^- = c_0^2 + \frac{(\gamma+1)^2}{8}u^2 - A^{-2} \qquad \text{so that}$$

$$B^- = c_0^2 + \frac{(\gamma+1)^2}{8}u^2 - \frac{(\gamma+1)^2}{8}\frac{u^2}{2} = c_0^2 + \frac{(\gamma+1)^2}{16}u^2$$

then

$$D^- = A^- + \sqrt{B^-} = \frac{(\gamma+1)}{4}u + \sqrt{c_0^2 + \frac{(\gamma+1)^2}{16}u^2}$$

$$D^+ = A^+ + \sqrt{B^+} = \sqrt{c_0^2 + \frac{(\gamma+1)^2}{16}u^2} + \frac{(\gamma+1)}{4}u$$

so that $D^+ = D^-$ and finally,

$$\boxed{D = \frac{(\gamma+1)}{4}u + \sqrt{c_0^2 + \left(\frac{\gamma+1}{4}\right)^2 u^2}} \tag{1.38}$$

This is the ideal gas Hugoniot in the D, u plane.

Now consider a monatomic ideal gas with $\gamma = 5/3$.

$$\frac{(\gamma+1)}{4} = \frac{2}{3}, \quad (\gamma-1) = \frac{2}{3}, \quad \text{and} \quad (\gamma+1) = \frac{8}{3}$$

Substituting this in relations (1.28), (1.31), (1.32), (1.33), (1.34), (1.35), and (1.38) gives

$$\frac{\rho}{\rho_0} = \left(\frac{4S+1}{S+4}\right) \tag{1.39}$$

$$u = \frac{3(S-1)}{\sqrt{5(4S+1)}} c_0 \tag{1.40}$$

$$D = \sqrt{\frac{(4S+1)}{5}} c_0 \tag{1.41}$$

$$c = \sqrt{\frac{S(S+4)}{4S+1}} c_0 \tag{1.42}$$

$$P = \frac{4V_0 - V}{4V - V_0} P_0 = S P_0 \tag{1.43}$$

$$b = \frac{S+4}{5} c_0 \tag{1.44}$$

$$D = \frac{2}{3}u + \sqrt{c_0^2 + \left(\frac{2u}{3}\right)^2} \tag{1.45}$$

The limiting compression is seen to be

$$\left(\frac{\rho}{\rho_0}\right)_L = \frac{\gamma+1}{\gamma-1} = \frac{5/3+1}{5/3-1} = 4$$

Now from relation (1.26) we have

$$c^2 = \left(\frac{T}{T_0}\right) c_0^2$$

and from relation (1.42) we have

$$c^2 = \frac{S(S+4)}{4S+1} c_0^2$$

so that

$$\left(\frac{T}{T_0}\right) = \frac{S(S+4)}{4S+1}$$

Solving for S, we obtain

$$S = 2\left(\frac{T}{T_0} - 1\right) + \sqrt{4\left(\frac{T}{T_0} - 1\right)^2 + \frac{T}{T_0}}$$

Let $T_0 = 300$ K. Beyond $T \cong 3000$ K ionization becomes important, so the limiting value of S for the model to be useful is determined by $T/T_0 \cong 10$.

$$S = 18 + \sqrt{334} = 36.28$$

The model is thus good for

$$\frac{P}{P_0} = S \leq 36, \quad \frac{\rho}{\rho_0} \leq 3.63, \quad D \leq 5.4160,$$
$$u \leq 3.91\, c_0, \quad c \leq 3.16\, c_0$$

From relation (1.44) we have $b \leq 8\, c_0$ so that $b/c \leq 2.55$ and thus, using the Hugoniot in place of the isentrope for such a strong shock would give a very poor approximation for the sound speed.

An overtaking rarefaction would move at the velocity c relative to the material behind the shock. In lab frame, we must add the material velocity u, so that the velocity of an overtaking rarefaction in lab frame is $u + c$. The velocity of the shock front in lab frame is D, however. From relations (1.40-1.42), we have

$$\frac{u+c}{D} = \frac{3(S-1) + \sqrt{5S(S+4)}}{(4S+1)}$$

We thus have $(u + c) > D$ for all shock strengths $S > 1$, and $D > c_0$. The shock is thus supersonic relative to the undisturbed medium, but subsonic relative to the material behind it. Rarefactions from the rear can thus overtake the shock and weaken it.

1.6 Simple Collision Analysis

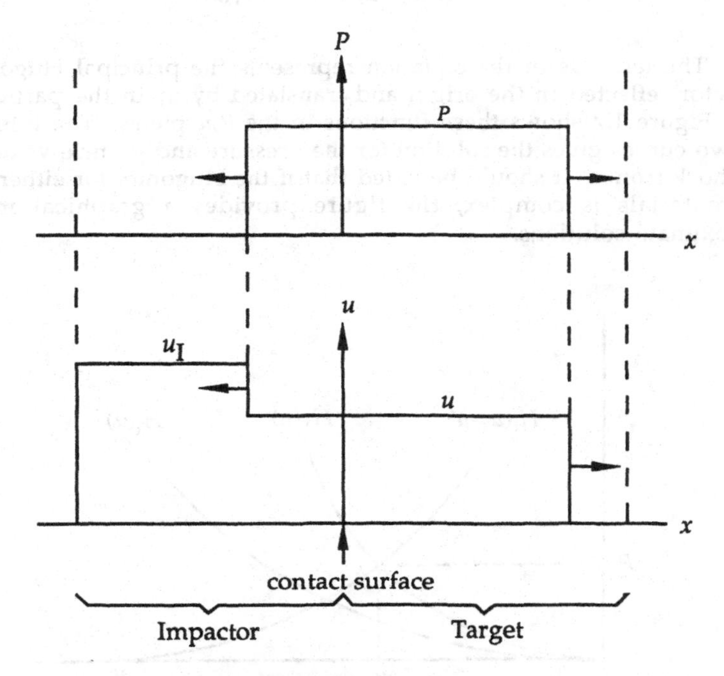

Fig. 1.6. Pressure and particle velocity distributions shortly after impact of an impactor with initial velocity u_I into an initially stationary target.

Consider the collision of an impactor with a stationary target, not necessarily of the same material. Pressure and particle velocity must be continuous across the contact surface if a gap doesn't open. Figure 1.6 illustrates the pressure and particle distributions after a short time. The places where the pressure drops to zero are the shock fronts in the two materials. Note that if the two materials are different, the shock velocities in the materials will be different. For the target, the jump in particle velocity is u, while for the impactor it is (u_I - u.)

Let H_I represent the principal Hugoniot of the impactor in the P,u plane, and H_T the corresponding principal Hugoniot for the target. Continuity of pressure across the contact surface requires

$$H_I(u_I - u) = P = H_T(u)$$

The left side of the equation represents the principal Hugoniot of the impactor reflected in the origin and translated by u_I in the particle velocity axis. Figure 1.7 shows these functions in the P,u plane. The intersection of the two curves gives the solution for the pressure and particle velocity behind the shock fronts. It should be noted that if the Hugoniot for either or both of the materials is complex, the figure provides a graphical method for approximate solutions.

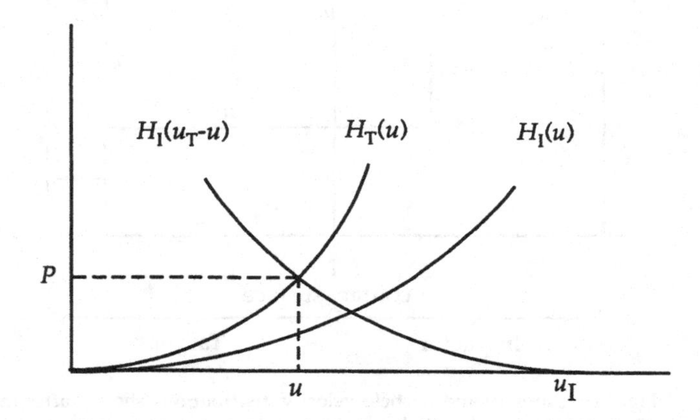

Fig. 1.7. Simple collision analysis in the P, u plane. $H_I(u)$ and $H_T(u)$ are the principal Hugoniots for the impactor and target, respectively. The intersection of the reflected and translated impactor Hugoniot with the target Hugoniot gives the solution for the pressure and particle velocities behind the shock.

For the case of symmetric impact, where the target and impactor are of the same material, we have $H_I(u) = H_T(u)$ and thus the solution for u and P is given by $u = u_I/2$, which can be substituted in the Hugoniot to obtain the pressure. For symmetric collisions therefore, one need only measure the impactor velocity u_I, and the shock velocity D along with the initial density ρ_0 to determine a point of the Hugoniot since conservation of momentum gives $P = \rho_0 D u_I / 2$.

2. Hugoniot Relations for Shock Waves with Curved Surfaces

2.1 Cylindrical Shock Waves

The Hugoniot equations are valid across a smoothly curved shock surface if $\rho_0 D$ is interpreted as the mass of fluid swept over by the shock front per unit area per unit time, where D is the shock velocity in lab frame, ρ_0 is the density of the undisturbed fluid, and the mass velocities in the equations are interpreted as the components of the fluid velocity normal to the shock surface[5]. As components of the fluid velocity parallel to the shock surface are continous across the shock, the result can be a change in the direction of fluid flow across the surface. The case of a cylindrical shock, such as is produced by an exploding wire is of particular interest.

Consider a shock of radius R moving radially outward into an undisturbed fluid at rest. In time δt the mass in a cylindrical shell between the shock and a surface of radius $R + D\delta t$ will flow into the shock, and a mass contained between the shock and a surface of radius $R\text{-}(D\text{-}u)\delta t$ will flow out of it, where u is the fluid velocity behind the shock in the laboratory frame. Let M_0 denote the mass per unit length in the shell ahead of the shock and M_i the value for the shell behind it. We have

$$M_0 = \pi\rho_0\left[(R+D\delta t)^2 - R^2\right]$$

and

$$M_i = \pi\rho\left\{R^2 - \left[R-(D-u)\delta t\right]^2\right\}$$

Conservation of mass then gives

$$\rho_0\left[(R+D\delta t)^2 - R^2\right] = \rho\left\{R^2 - \left[R-(D-u)\delta t\right]^2\right\}$$

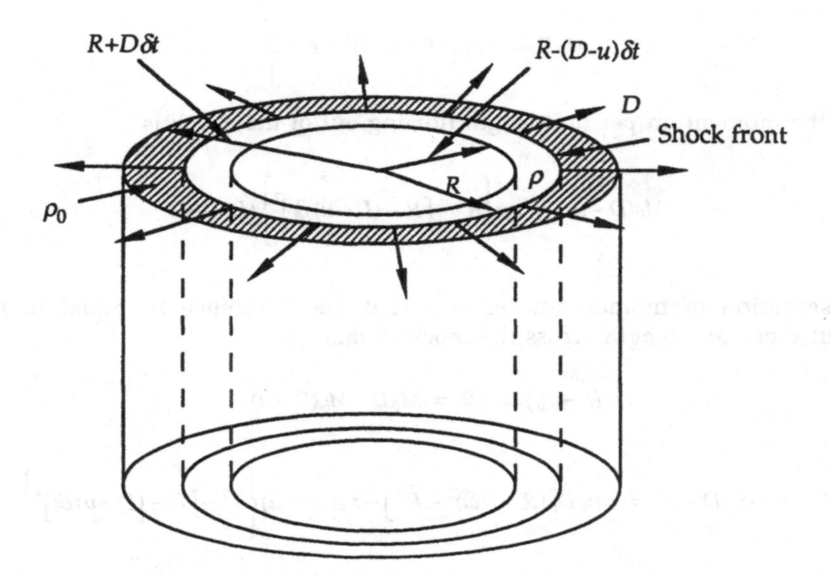

Fig. 2.1. Expanding cylindrical shock wave with velocity D. A unit length of the cylindrical wave is shown. ρ is the density of the shocked medium and ρ_0 is the initial density of the undisturbed medium at rest. u is the mass velocity.

Solve for ρ/ρ_0 and expand the products in the numerator and denominator to obtain

$$\frac{\rho}{\rho_0} = \frac{D}{(D-u)} \frac{2R + D\delta t}{\left[2R - (D-u)\delta t\right]}$$

If we choose $D\delta t$ of the order of the shock thickness, we can easily satisfy
$R >> D\delta t$ so that

$$\frac{\rho}{\rho_0} \longrightarrow \frac{D}{(D-u)}$$

which is the usual expression for conservation of mass. The momentum per unit length flowing into the shock is

$$M_0 D = \pi\rho_0\left[(R+D\delta t)^2 - R^2\right]D$$

and the momentum per unit length flowing out of the shock is

$$M_i(D-u) = \pi\rho\left\{R^2 - \left[R-(D-u)\delta t\right]^2\right\}(D-u)$$

Conservation of momentum requires that the difference be equal to the impulse per unit length across the shock so that

$$(P-P_0)2\pi R\,\delta t = M_0 D - M_i(D-u)$$

Now

$$M_o D - M_i(D-u) = \pi\rho_0 D\left[(R+D\delta t)^2 - R^2\right] - \pi\rho(D-u)\left\{R^2 - \left[R-(D-u)\delta t\right]^2\right\}$$

but

$$\rho = \rho_0 \frac{\left[(R+D\delta t)^2 - R^2\right]}{\left\{R^2 - \left[R-(D-u)\delta t\right]^2\right\}}$$

so that substitution gives

$$M_o D - M_i(D-u) = \pi\rho_0 D\left[(R+D\delta t)^2 - R^2\right]$$

$$-\pi\rho_0 \frac{(D-u)\left\{R^2 - \left[R-(D-u)\delta t\right]^2\right\}\left[(R+\delta t)^2 - R^2\right]}{\left\{R^2 - \left[R-(D-u)\delta t\right]^2\right\}}$$

$$= \pi\rho_0\left[(R+D\delta t)^2 - R^2\right](D-D+u) = \pi\rho_0 u\left[(R+D\delta t)^2 - R^2\right]$$

We then have

$$P - P_0 = \frac{\pi\rho_0 u\left[(R + D\delta t)^2 - R^2\right]}{2\pi R\,\delta t} = \rho_0 Du\left[1 + \frac{D\delta t}{R}\right]$$

For $R >> D\delta t$, we have $P\text{-}P_0 \rightarrow \rho_0 Du$, which is the usual expression for momentum conservation. The derivation for the change in internal energy is exactly as before.

A plane shock wave moves at constant velocity (unless overtaken by a rarefaction), but in a curved shock front the velocity of the wave must continually change. The Hugoniot equations remain valid however, and the observed shock velocity at any point may be used with the principal Hugoniot of the fluid to determine the pressure behind the wave relative to that ahead of it.

2.2 Spherical Shock Waves

The analysis for the spherical case uses spherical shells instead of cylindrical ones, so that using the same nomenclature, we have

$$M_o = \frac{4\pi}{3}\rho_0\left[(R + D\delta t)^3 - R^3\right]$$

$$M_i = \frac{4\pi}{3}\rho\left\{R^3 - \left[R - (D-u)\delta t\right]^3\right\}$$

and

$$(P - P_0)4\pi R^2\,\delta t = M_o D - M_i (D - u)$$

Conservation of mass gives

$$\rho_0\left[(R + D\delta t)^3 - R^3\right] = \rho\left\{R^3 - \left[R - (D-u)\delta t\right]^3\right\}$$

so that

$$\frac{\rho}{\rho_0} = \frac{\left[(R+D\delta t)^3 - R^3\right]}{\left\{R^3 - \left[R - (D-u)\delta t\right]^3\right\}}$$

Expand the numerator and denominator and then factor $D\delta t$ out of the numerator and $(D\text{-}u)\delta t$ out of the denominator to obtain

$$\frac{\rho}{\rho_0} = \frac{D}{(D-u)}\left[\frac{3R^2 + 3RD\delta t + (D\delta t)^2}{3R^2 - 3R(D-u)\delta t + (D-u)^2(\delta t)^2}\right]$$

Now since $D > u$, for $R >> D\delta t$, we also have $R >> (D\text{-}u)\delta t$ and

$$\frac{\rho}{\rho_0} \rightarrow \frac{D}{(D-u)}$$

Now substituting M_o and M_i and collecting terms we obtain

$$M_o D - M_i (D-u) = \frac{4\pi}{3}\rho_0 u\left[(R+D\delta t)^3 - R^3\right]$$

then

$$P - P_0 = \frac{4\pi\rho_0 u}{12\pi R^2 \delta t}\left[(R+D\delta t)^3 - R^3\right] = \frac{\rho_0 u}{3R^2 \delta t}\left[(R+D\delta t)^3 - R^3\right]$$

but

$$\left[(R+D\delta t)^3 - R^3\right] = D\delta t\left[3R^2 + 3RD\delta t + (D\delta t)^2\right]$$

so that

$$P - P_0 = \rho_0 D u\left[1 + \frac{D\delta t}{R} + \frac{1}{3}\left(\frac{D\delta t}{R}\right)^2\right]$$

and thus, for $R >> D\delta t$ we have

$$P - P_0 \rightarrow \rho_0 D u$$

so that the Hugoniot relations for conservation of mass and momentum again take the same form as for plane shock waves.

3. Stability of Shock Waves

3.1 Stability of Shocks as a Function of Hugoniot Shape

If a shock were supersonic relative to the medium behind it, it would become "uncontrollable", since it would not be influenced by any process occurring behind it. That is one reason that a rarefaction shock is impossible in materials with normal Hugoniots. For most materials, the compressibility decreases as the pressure increases. As a result, the Hugoniot will be concave upward in the P, V plane. Consider such a Hugoniot illustrated in figure 3.1. The tangent to the Hugoniot is shown at the final state and subtends angle θ_1 relative to the volume axis, while the Rayleigh line subtends angle θ_2. For a Hugoniot that is concave upward, $\theta_1 > \theta_2$. We have

$$\tan\theta_1 = -\left(\frac{\partial P}{\partial V}\right)_H, \quad \tan\theta_2 = \left(\frac{P-P_0}{V_0-V}\right)$$

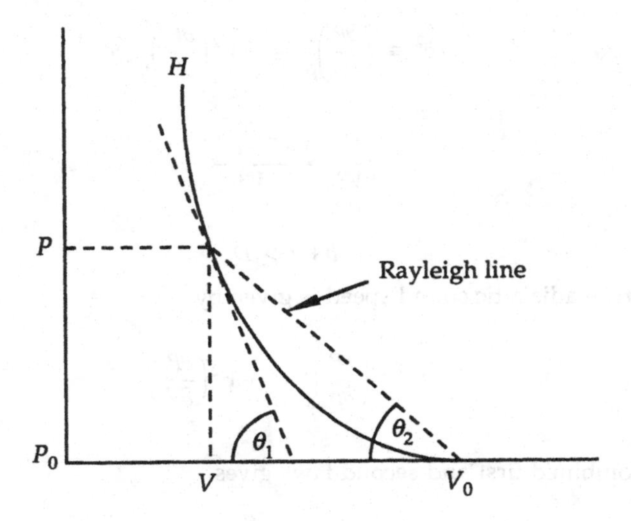

Fig. 3.1. Hugoniot with curvature concave upward in the P, V plane.

If the Hugoniot is concave upward, as shown in the figure, then

$$-\left(\frac{\partial P}{\partial V}\right)_H > \left(\frac{P-P_0}{V_0-V}\right)$$

Now from the Hugoniot equations for a medium initially at rest,

$$D = V_0\sqrt{\frac{P-P_0}{V_0-V}} \quad , \quad u = \sqrt{(P-P_0)(V_0-V)}$$

Thus, after algebraic manipulation, we can write

$$(D-u)^2 = V^2\left(\frac{P-P_0}{V_0-V}\right)$$

so that we have

$$-\left(\frac{\partial P}{\partial V}\right)_H > \frac{(D-u)^2}{V^2}$$

but

$$b^2 \equiv \left(\frac{\partial P}{\partial \rho}\right)_H = -V^2\left(\frac{\partial P}{\partial V}\right)_H$$

so that

$$\frac{b^2}{V^2} > \frac{(D-u)^2}{V^2}$$

or

$$b+u > D$$

Now the true adiabatic sound speed is given by

$$c^2 \equiv \left(\frac{\partial P}{\partial \rho}\right)_S = -V^2\left(\frac{\partial P}{\partial V}\right)_S$$

and the combined first and second laws gives

$$TdS = dE + PdV$$

but also, on the Hugoniot

$$E-E_0 = \frac{1}{2}(P+P_0)(V_0-V)$$

so that

$$dE_H = \frac{1}{2}(V_0-V)dP_H - \frac{1}{2}(P+P_0)dV_H$$

Substituting gives

$$TdS_H = dE_H + PdV_H = \frac{1}{2}(V_0-V)dP_H + \frac{1}{2}(P-P_0)dV_H$$

Now

$$dP \equiv \left(\frac{\partial P}{\partial V}\right)_S dV + \left(\frac{\partial P}{\partial S}\right)_V dS \quad \text{and} \quad \left(\frac{\partial P}{\partial V}\right)_S = -\frac{c^2}{V^2}$$

so that

$$dP = -\frac{c^2}{V^2}dV + \left(\frac{\partial P}{\partial S}\right)_V dS$$

Now from thermodynamics

$$TdS \equiv \frac{\kappa_T c_v}{\beta}dP + \frac{c_P}{\beta V}dV, \text{ where } \beta = \frac{1}{V}\left(\frac{\partial V}{\partial T}\right)_P, \text{ and } \kappa_T = -\frac{1}{V}\left(\frac{\partial V}{\partial P}\right)_T$$

then

$$TdS_V = \frac{\kappa_T c_v}{\beta}dP_V \quad \text{so that} \quad \left(\frac{\partial P}{\partial S}\right)_V = \frac{\beta T}{\kappa_T c_v}$$

Substituting for β and κ_T we have

$$\left(\frac{\partial P}{\partial S}\right)_V = -\frac{T}{c_v}\left(\frac{\partial V}{\partial T}\right)_P\left(\frac{\partial P}{\partial V}\right)_T$$

but

$$-\left(\frac{\partial V}{\partial T}\right)_P\left(\frac{\partial P}{\partial V}\right)_T \equiv \left(\frac{\partial P}{\partial T}\right)_V$$

thus

$$\left(\frac{\partial P}{\partial S}\right)_V = \frac{T}{c_v}\left(\frac{\partial P}{\partial T}\right)_V \quad \text{and} \quad \frac{V}{T}\left(\frac{\partial P}{\partial S}\right)_V = \frac{V}{c_v}\left(\frac{\partial P}{\partial T}\right)_V$$

Now

$$dE_v = TdS_v \qquad \text{so that} \qquad \left(\frac{\partial}{\partial E}\right)_V = \frac{1}{T}\left(\frac{\partial}{\partial S}\right)_V$$

and

$$\gamma_G \equiv V\left(\frac{\partial P}{\partial E}\right)_V = \frac{V}{T}\left(\frac{\partial P}{\partial S}\right)_V$$

and

$$\left(\frac{\partial P}{\partial S}\right)_V = \frac{\gamma_G T}{V}$$

γ_G is called the Grüneisen coefficient.

We can thus write

$$dP = -\frac{c^2}{V^2}dV + \frac{T}{V}\gamma_G dS$$

or

$$dP_H = -\frac{c^2}{V^2}dV_H + \frac{\gamma_G}{V}TdS_H$$

$$dP_H = -\frac{c^2}{V^2}dV_H + \frac{\gamma_G}{V}\left[\frac{1}{2}(V_0 - V)dP_H + \frac{1}{2}(P - P_0)dV_H\right]$$

Collecting terms gives

$$\left[V - \frac{\gamma_G}{2}(V_0 - V)\right]dP_H = \left[\frac{\gamma_G}{2}(P - P_0) - \frac{c^2}{V}\right]dV_H$$

so that

$$\left(\frac{\partial P}{\partial V}\right)_H = \frac{1}{V^2}\frac{\frac{1}{2}\gamma_G V(P - P_0) - c^2}{\left[1 - \frac{\gamma_G}{2V}(V_0 - V)\right]}$$

We then have

$$b^2 = -V^2\left(\frac{\partial P}{\partial V}\right)_H = \frac{c^2 - \frac{1}{2}\gamma_G V(P - P_0)}{\left[1 - \frac{\gamma_G}{2V}(V_0 - V)\right]}$$

Now we have shown that

$$(D-u)^2 = V^2\left(\frac{P-P_0}{V_0-V}\right)$$

so that

$$V(P-P_0) = (D-u)^2\left(\frac{V_0}{V}-1\right) = (D-u)^2\left(\frac{\rho}{\rho_0}-1\right)$$

Substitution and manipulation gives

$$\left[c^2-(D-u)^2\right] = \left[b^2-(D-u)^2\right]\left[1-\frac{\gamma_G}{2}\left(\frac{\rho}{\rho_0}-1\right)\right]$$

but we have already shown that

$$\left[b^2-(D-u)^2\right] > 0$$

so that

$$\left[c^2-(D-u)^2\right] > 0 \quad \text{if} \quad \left[1-\frac{\gamma_G}{2}\left(\frac{\rho}{\rho_0}-1\right)\right] > 0$$

This reduces to

$$c+u > D \quad \text{if} \quad \frac{\rho}{\rho_0} < \left(\frac{2}{\gamma_G}+1\right)$$

Thus, for a Hugoniot that is concave upward, the shock will be subsonic relative to the medium behind it for compressions such that

$$\frac{\rho}{\rho_0} < \left(\frac{2}{\gamma_G}+1\right)$$

Now consider the ideal gas:

$$P = (\gamma-1)\frac{E}{V}$$

$$V\left(\frac{\partial P}{\partial E}\right)_V = \gamma-1 = \gamma_G, \quad \frac{2}{\gamma_G}+1 = \frac{\gamma+1}{\gamma-1}$$

This is just the limiting compression. The shock is thus always subsonic relative to the medium behind it for the ideal gas.

3.2 Influence of Phase Changes

Many phase changes produce a two-wave structure. If the shock velocity decreases with pressure, the shock can break up into two or more waves, or possibly one wave with a continuously smeared front. Consider the stability of a two-wave structure. Figure 3.2 illustrates the distribution of pressure and particle velocity.

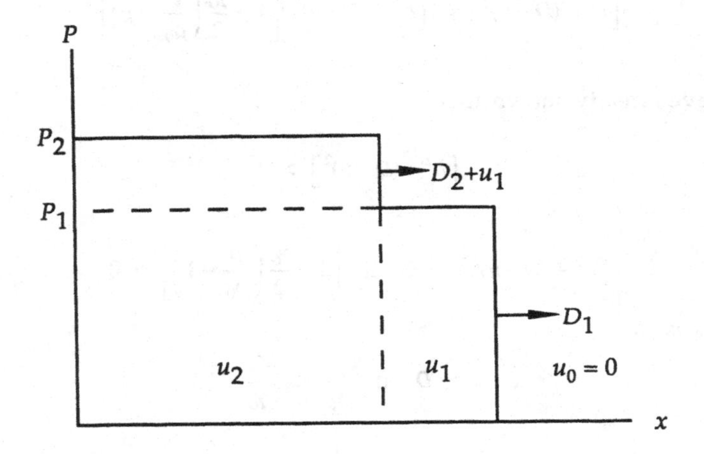

Fig. 3.2. Distribution of pressure and particle velocity behind the shock fronts in a two-wave structure caused by a phase change. u_1 is the particle velocity behind the first wave, and u_2 is the velocity behind the second wave. The material is at rest ahead of the first wave.

The initial shock moves with a velocity D_1 in the lab frame into material at rest. The particle velocity behind the front is u_1. A second, following shock which gives a jump to a higher pressure travels at velocity D_2 relative to the medium ahead of it, but at velocity D_2+u_1 relative to the lab frame, since the material ahead of the shock is already moving at velocity u_1. The particle velocity behind the second shock is u_2. For the first shock, with $u_0 = 0$, relation (1.6) gives

$$D_1 = V_0\sqrt{\frac{P_1 - P_0}{V_0 - V_1}}$$

where V is the specific volume, and relation (1.5) gives

$$u_1 = \sqrt{(P_1 - P_0)(V_0 - V_1)}$$

The lab frame velocity for the second shock is

$$D_2 = u_1 + V_1\sqrt{\frac{P_2 - P_1}{V_1 - V_2}}$$

and relation (1.5) gives

$$u_2 = u_1 + \sqrt{(P_2 - P_1)(V_1 - V_2)}$$

Now if

$$D_2 \geq D_1$$

then the second shock will overtake the first and merge with it. This is the condition for a stable shock. Substituting for D_2, u_1, and D_1, we have

$$V_1\sqrt{\frac{P_2 - P_1}{V_1 - V_2}} + \sqrt{(P_1 - P_0)(V_0 - V_1)} \geq V_0\sqrt{\frac{P_1 - P_0}{V_0 - V_1}}$$

Transferring the second term on the left to the right side and factoring, gives

$$V_1\sqrt{\frac{P_2 - P_1}{V_1 - V_2}} \geq \sqrt{P_1 - P_0}\left[\frac{V_0}{\sqrt{V_0 - V_1}} - \frac{V_0 - V_1}{\sqrt{V_0 - V_1}}\right]$$

or

$$V_1\sqrt{\frac{P_2-P_1}{V_1-V_2}} \geq V_1\sqrt{\frac{P_1-P_0}{V_0-V_1}}$$

or, finally,

$$\boxed{\frac{P_2-P_1}{V_1-V_2} \geq \frac{P_1-P_0}{V_0-V_1}} \tag{3.1}$$

Now consider a Hugoniot with normal curvature, but a cusp in it. Figure 3.3 illustrates the situation. The shock forms by first establishing the state at point 1 and then establishing the state at point 2. The stability condition reduces to $\alpha_2 \geq \alpha_1$. If the slope of the second Rayleigh line R_2 is the same or steeper than that of the first, R_1, then a single wave structure results. For final states between points A and B a two-wave structure develops, with the second wave falling behind the first. If one is making shock wave measurements with shorting-pin detectors, the first wave, giving pressure P_1 will close the circuit (unless it is a very weak wave) and the second wave will not be seen. As the shock velocity is increased on successive experiments, point B will eventually be reached and the final pressure P_2 will be indicated because the two waves merge. Until this occurs, the particle velocity appears to increase without any corresponding increase in the measured shock velocity. After the point B is passed the D, u plot appears to have a new segment. Figure 3.4 illustrates the situation. The extension of the second segment between points 2 and B is not seen in the measurements. The flat region between points 1 and 2 does not necessarily represent equation of state (EOS) data. At point 1 the particle and shock velocities are meaningful and can be used to determine the pressure of the first wave. If equilibrium conditions exist, it is the pressure where the cusp is located. If the cusp is the result of a phase change, this is the pressure of the phase change. An example is iron at 130 kBar. Now let's examine the details in the phase change that creates a cusp in the Hugoniot. Figure 3.5 illustrates the Hugoniot along with the coexistence region for the two phases.

First consider the isothermal process indicated by the dotted curve. The pressure increases along the isotherm until the phase line is reached at V_a. The finite volume reduction V_a - V_b occurs at constant temperature

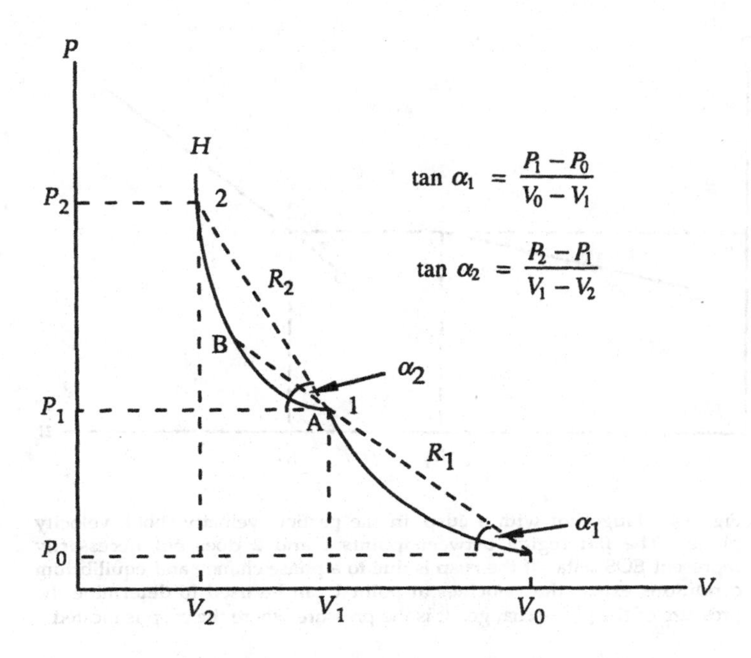

Fig. 3.3. Stability analysis for a Hugoniot with a cusp due to a phase change. R_1 is the Rayleigh line for the first jump and R_2 is the Rayleigh line for the second jump. For waves too weak to reach point A, or strong enough that point 2 is above point B, a single wave results. For waves that would fall between points A and B, a two-wave structure results because α_2 is less than α_1 so that the second jump gives a wave with slower velocity.

and pressure, accompanied by the enthalpy change L, corresponding to the latent heat of the phase change. When volume V_b is reached, pressure continues to increase in phase b for further volume reductions.

Now consider a shock wave connecting state P_0, V_0 to the state P_1, V_1 on the phase line, and a second (assumed single) shock connecting P_1, V_1 to the state P_2, V_2 (note that for the case shown, a two-wave structure will result.) We have

$$h_1 = E_1 + P_1 V_1, \qquad h_2 = E_2 + P_2 V_2$$

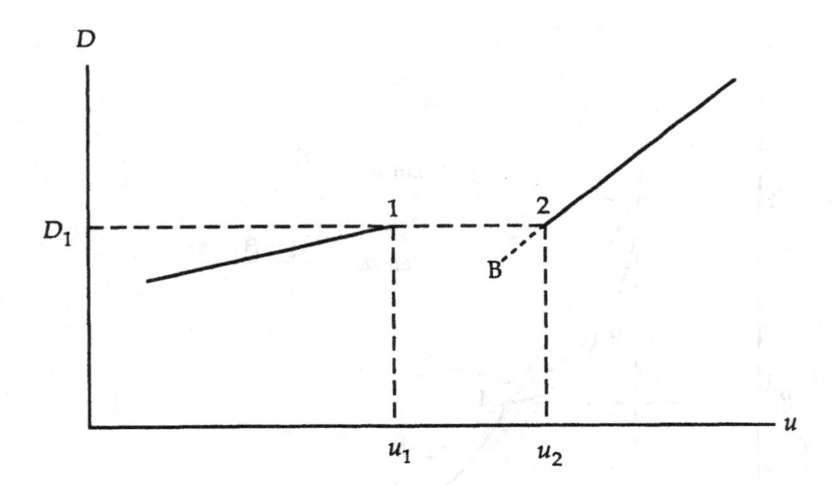

Fig. 3.4. Hugoniot with a cusp, in the particle velocity-shock velocity plane. The flat region between points 1 and 2 does not necessarily represent EOS data. If the cusp is due to a phase change and equilibrium conditions exist, the velocities at point 1 can be used to determine the pressure of the phase change. It is the pressure where the cusp is located.

The Hugoniot energy conservation relation gives

$$E_1 - E_0 = \frac{1}{2}(P_1 + P_0)(V_0 - V_1)$$

for the single shock connecting P_0, V_0 and P_1, V_1, and

$$E_2 - E_1 = \frac{1}{2}(P_2 + P_1)(V_1 - V_2)$$

for a single shock connecting P_1, V_1 and P_2, V_2. Now calculate the enthalpy change from P_1, V_1 to P_2, V_2.

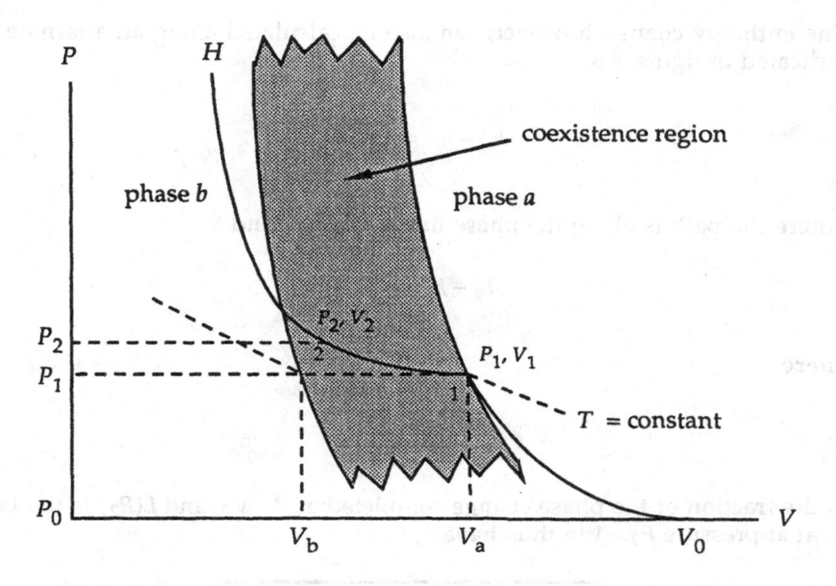

Fig. 3.5. Schematic representation of a Hugoniot passing through a phase change. The solid curve denoted H is the Hugoniot, with a cusp at point 1. An isotherm through the point is shown by the dotted curve.

$$h_2 - h_1 = E_2 - E_1 + P_2 V_2 - P_1 V_1 = \frac{1}{2}(P_2 + P_1)(V_1 - V_2) + P_2 V_2 - P_1 V_1$$

$$= \frac{1}{2} P_2 (V_1 - V_2 + 2V_2) + \frac{1}{2} P_1 (V_1 - V_2 - 2V_1)$$

$$= \frac{1}{2} P_2 (V_1 + V_2) - \frac{1}{2} P_1 (V_1 + V_2) = \frac{1}{2}(P_2 - P_1)(V_1 + V_2)$$

so that

$$\boxed{h_2 - h_1 = \frac{1}{2}(P_2 - P_1)(V_1 + V_2)} \tag{3.2}$$

The enthalpy change however, can also be calculated along an alternate path, indicated in figure 3.6.

$$h_{a_2} - h_1 = \left[\int_{P_1}^{P_2} dh\right]_a$$

where the path is along the phase line as shown, and

$$h_2 - h_{a_2} = \lambda L(P_2)$$

where

$$\lambda = \frac{V_{a_2} - V_2}{V_{a_2} - V_{b_2}}$$

is the fraction of the phase change completed at P_2, V_2, and $L(P_2)$ is the latent heat at pressure P_2. We thus have

$$\boxed{\begin{aligned} h_2 - h_1 &= \left[\int_{P_1}^{P_2} dh\right]_a + \lambda L(P_2) \\ \text{where} \quad \lambda &= \frac{V_{a_2} - V_2}{V_{a_2} - V_{b_2}} \end{aligned}} \tag{3.3}$$

Now let $P_{ph}(T)$ be the coexistence curve in the P, T plane. The Clausius-Clapeyron equation then gives

$$\boxed{\frac{dP_{ph}(T)}{dT} = \frac{-L}{T(V_a - V_b)}} \tag{3.4}$$

$$L = -T\left(\frac{dP_{ph}}{dT}\right)(V_a - V_b)$$

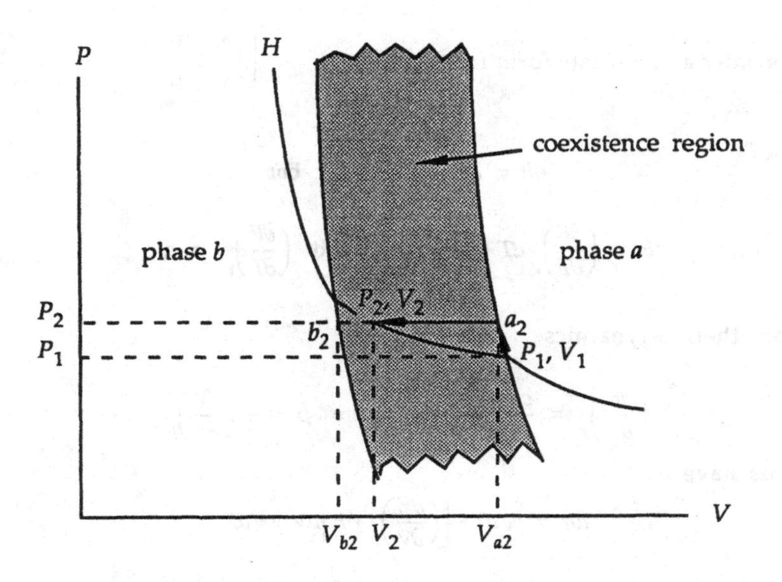

Fig. 3.6. Alternate path for calculation of h_1-h_2 along the phase line from P_1, V_1 to a_2 and then along an isobar from a_2 to P_2, V_2.

so that

$$L(P_2) = -T\left(\frac{dP_{ph}}{dT}\right)_{P_2}(V_{a_2} - V_{b_2})$$

and

$$\lambda L(P_2) = -T\left(\frac{dP_{ph}}{dT}\right)_{P_2}\frac{(V_{a_2} - V_{b_2})(V_{a_2} - V_2)}{(V_{a_2} - V_{b_2})}$$

and thus

$$\boxed{\lambda L(P_2) = -T\left(\frac{dP_{ph}}{dT}\right)_{P_2}(V_{a_2} - V_2)} \tag{3.5}$$

Now consider an alternate form for the integral $\left[\int_{P_1}^{P_2} dh\right]_a$

$$h = E + PV$$

$$dh = dE + VdP + PdV \quad \text{but}$$

$$dE = \left(\frac{\partial E}{\partial T}\right)_V dT + \left(\frac{\partial E}{\partial V}\right)_T dV \quad \text{and} \quad \left(\frac{\partial E}{\partial T}\right)_V = c_v,$$

and from thermodynamics

$$\left(\frac{\partial E}{\partial V}\right)_T = \frac{c_P - c_v}{\beta V} - P, \quad \text{where} \quad \beta = \frac{1}{V}\left(\frac{\partial V}{\partial T}\right)_P$$

We thus have

$$dh = c_V dT + \left[\left(\frac{\partial E}{\partial V}\right)_T + P\right] dV + VdP$$

or

$$dh = c_v dT + \left[\frac{c_P - c_v}{\beta V}\right] dV + VdP$$

but also, from thermodynamics we have

$$c_P - c_V = \frac{\beta^2 TV}{\kappa_T} \quad \text{where} \quad \kappa_T = -\frac{1}{V}\left(\frac{\partial V}{\partial P}\right)_T$$

so that

$$\frac{c_P - c_V}{\beta V} = \frac{\beta T}{\kappa_T} = \frac{(1/V)(\partial V/\partial T)_P}{(-1/V)(\partial V/\partial P)_T} T = -T\left(\frac{\partial V}{\partial T}\right)_P \left(\frac{\partial P}{\partial V}\right)_T$$

and since

$$\left(\frac{\partial V}{\partial T}\right)_P \left(\frac{\partial P}{\partial V}\right)_T \left(\frac{\partial T}{\partial P}\right)_V \equiv -1$$

we can write

$$-\left(\frac{\partial V}{\partial T}\right)_P\left(\frac{\partial P}{\partial V}\right)_T = \left(\frac{\partial P}{\partial T}\right)_V$$

so that

$$\frac{c_P - c_V}{\beta V} = \frac{\beta T}{\kappa_T} = T\left(\frac{\partial P}{\partial T}\right)_V$$

Substitution then gives

$$\boxed{dh = c_V dT + T\left(\frac{\partial P}{\partial T}\right)_V dV + VdP} \qquad (3.6)$$

This can then be integrated to give $\left[\int_{P_1}^{P_2} dh\right]_a$.

From relation (3.2) we have

$$\frac{1}{2}(P_2 - P_1)(V_1 + V_2) = \left[\int_{P_1}^{P_2} dh\right]_a + \lambda L(P_2),$$

but

$$\left[\int_{P_1}^{P_2} dh\right]_a = \left[\int_{T_1}^{T_2} c_v\, dT + \int_{V_1}^{V_{a_2}} T\left(\frac{\partial P}{\partial T}\right)_v dV + \int_{P_1}^{P_2} VdP\right]_a$$

so that we can write

$$\boxed{\begin{aligned}\frac{1}{2}(P_2 - P_1)(V_1 + V_2) = &\left[\int_{T_1}^{T_2} c_v\, dT + \int_{V_1}^{V_{a_2}} T\left(\frac{\partial P}{\partial T}\right)_V dV + \int_{P_1}^{P_2} VdP\right]_a \\ &- T\left(\frac{dP_{ph}}{dT}\right)_{P_2}(V_{a_2} - V_2)\end{aligned}} \qquad (3.7)$$

Thermodynamic data for the phase line $P_{ph}(T)$ and the quantities $V_a(P)$, c_v, and $(\partial P/\partial T)_v$ along the phase line allow calculation of the locus of points P_2, V_2 from the above expression. The boundary $V_b(P)$ of the more dense phase must be known to prevent using the expression beyond the region where it is valid. This is equivalent to knowing the latent heat, since V_b can be calculated from the Clausius-Clapeyron equation with the above data. We thus see that with the above data we can calculate a predicted Hugoniot through the coexistence region.

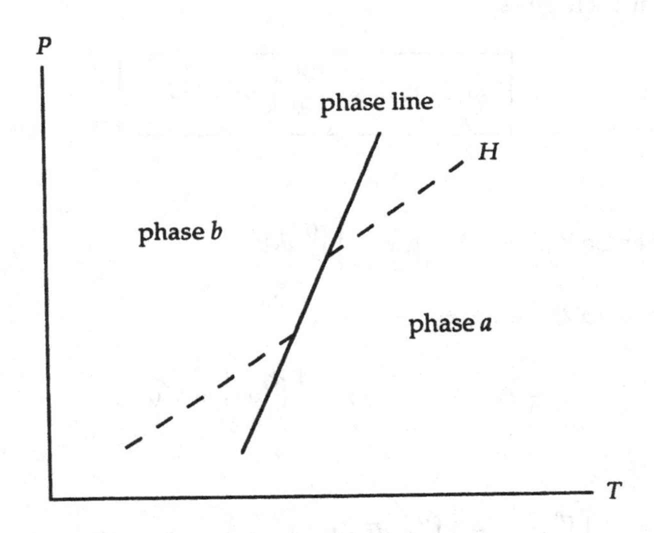

Fig. 3.7. Hugoniot crossing a phase line where the temperature increases more rapidly along the Hugnoiot than along the phase line. H is the Hugoniot.

Now consider some of the situations that may occur in first order phase changes (eg; $dP/dT > 0$ along the phase line.) Two different situations can occur:

1. The temperature increases more rapidly along the Hugoniot than along the phase line. Figure 3.7 illustrates the situation. One can shock from the high density phase b to the low density phase a (eg; melting, vaporization). Note that dP/dT is steeper on the phase line than on the Hugoniot.

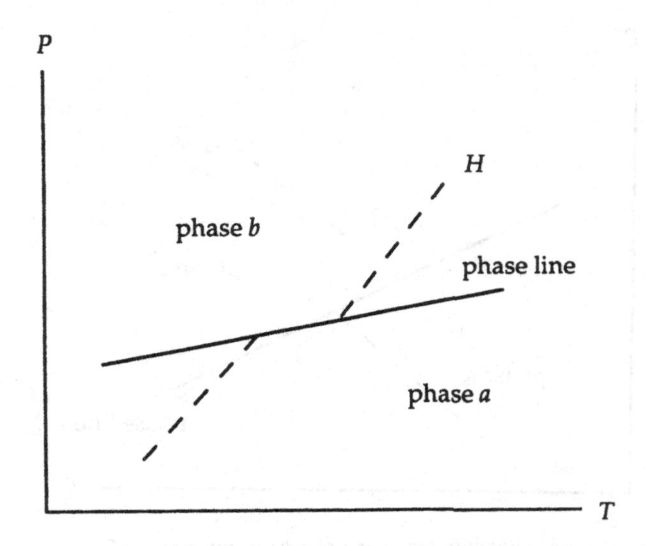

Fig. 3.8. Hugoniot crossing a phase line where the temperature increases more rapidly along the phase line than along the Hugoniot. H is the Hugoniot.

2. The temperature increases more rapidly along the phase line than along the Hugoniot (dP/dT is steeper on the Hugoniot than on the phase line.) Figure 3.8 illustrates this case. One can shock from the low density phase a to the high density phase (eg; solid-solid phase transformations). The Hugoniot has its initial state in the high temperature phase, and a two-wave structure will result (the mixed phase causes a dog-leg in the Hugoniot.) In either case 1 or case 2, in general, the Hugoniot may or may not intersect the phase line.

Now consider the case where $dP/dT < 0$ along the phase line. This case can correspond to anomalous melting or crystallographic changes. Figure 3.9 illustrates the situation. If the phase change exists, the Hugoniot must eventually cross it, provided the initial state is in the low temperature phase. Note that one cannot readily distinguish the above cases from the P, V or D, u plots. They have identical character. When the material is shocked from a low density phase to a high density phase there can be a noticeable effect on the Hugoniot in the D, u plane. In general, the larger the volume change the longer the flat central region in the D, u plot.

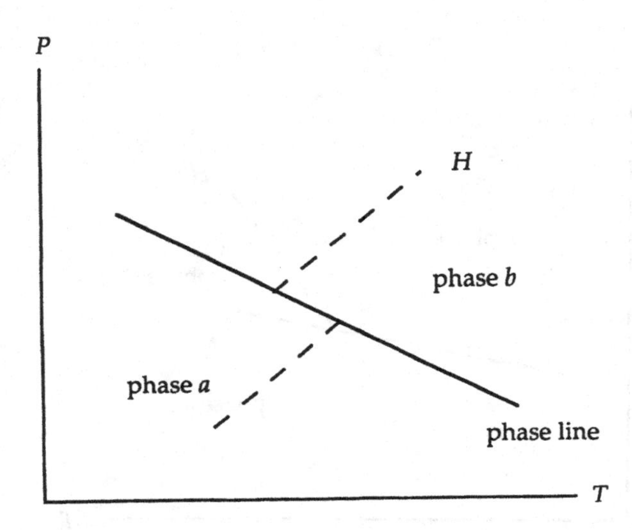

Fig. 3.9. Hugoinot crossing a phase line where $dP/dT < 0$ along the phase line. H is the Hugoniot.

During the shock through a phase transition, there can be a relaxation time involved, resulting in an "overdriven" Hugoniot curve. The first phase persists past the transition point and then relaxes to the Hugoniot for the higher pressure phase. The larger the overdriven pressure, the more the Hugoniot will be overdriven, by an amount corresponding to the original shock strength. The relaxation could be either time and/or pressure dependent. Figure 3.10 illustrates the situation.

Alternatively, the transformation could begin immediately but fail to go to completion, as shown in figure 3.11. The D, u plot is the same for either case, as shown in Figure 3.12. The choice of which case exists may be resolved by looking to see whether a two-wave structure exists. The data in the mixed phase region are far from an equilibrium state. It should be noticed that the one-dimensional nature of the experiment also gives rise to a two-wave structure. For the relatively low pressure shocks, the first wave is due to deformation in one dimension only, and hence propagates at the longitudinal sound velocity (elastic precursor). When the yield point is

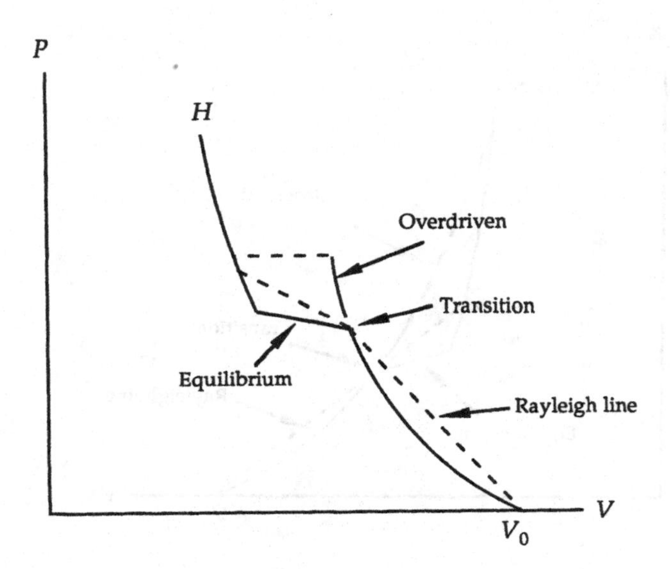

Fig. 3.10. Overdriven Hugoniot through a phase change caused by relaxation time effects. H is the normal Hugoniot.

reached, plastic behavior begins and the material relaxes toward its hydrodynamic equilibrium state.

Often, the cause of a two-wave structure as elastic-plastic or a phase transition can be resolved by examining the high pressure data. If these extrapolate to a value below the bulk sound speed, it must be concluded that a transformation exists. If the extrapolation is above the sound speed it could still be due to another type of transition (2nd order) where the compressibility changes. If the flat segment in the D, u curve does not occur near the measured longitudinal wave velocity, elastic-plastic behavior is ruled out. Two necessary conditions for establishing that the two-wave structure is due to an elastic-plastic wave are that the low pressure data extrapolate to the longitudinal sound speed or that the flat region is at or slightly higher than the longitudinal sound speed, and also the high pressure region extrapolates reasonably well to the bulk sound speed.

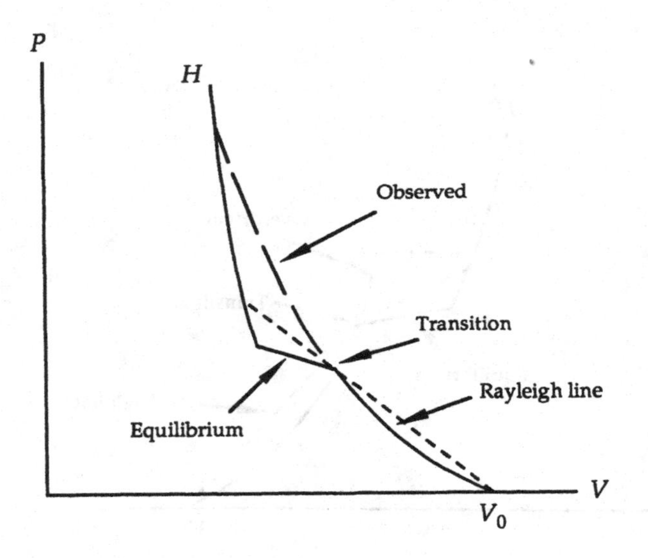

Fig. 3.11. Overdriven Hugoniot through a phase transition in which the transition fails to go to completion.

3.3 Influence of Material Strength

For an isotropic solid with sufficiently small strains we can write

$$\tau_{ij} = 2\mu e_{ij} + \lambda(e_{11} + e_{22} + e_{33}) \tag{3.8}$$

where μ, λ are the Lamé elastic constants, τ is the stress tensor and e the strain tensor.

1. For hydrostatic compression, we have

$$e_{11} = e_{22} = e_{33} = \frac{1}{3}\frac{\Delta V}{V}$$

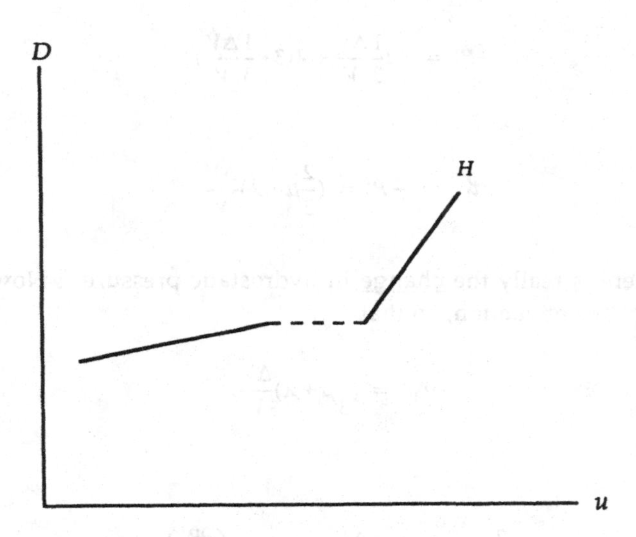

Fig. 3.12. D, u plot for the case of an overdriven shock through a phase transition. H is the apparent Hugoniot.

and off diagonal components vanish:

$$e_{ij} = \frac{1}{3}\frac{\Delta V}{V}\delta_{ij}$$

where δ_{ij} is the Kronecker delta operator. There is no shear stress, so τ is diagonal and the diagonal elements are equal to the negative of the pressure:

$$\tau'_{ij} = -P'\delta_{ij}$$

where the prime denotes the hydrostatic case. Substituting in relation (3.8) gives

$$\tau'_{11} = 2\mu e_{11} + \lambda(e_{11} + e_{22} + e_{33})$$

or

$$-P' = 2\mu\frac{1}{3}\frac{\Delta V}{V} + \lambda(3 \cdot \frac{1}{3}\frac{\Delta V}{V})$$

or

$$\tau'_{11} = -P' = (\frac{2}{3}\mu + \lambda)\frac{\Delta V}{V}$$

Note that P here is really the <u>change</u> in hydrostatic pressure. Now $\mu, \lambda > 0$ and $\tau'_{ii} = -P$ by convention, so that

$$-\Delta P = (\frac{2}{3}\mu + \lambda)\frac{\Delta V}{V}$$

and thus

$$(\frac{2}{3}\mu + \lambda) = -\frac{\Delta P'}{\Delta V}V \equiv -V\left(\frac{\partial P'}{\partial V}\right)$$

is the bulk modulus.

2. Now consider one dimensional compression in the x direction. By definition $e_{11} = \Delta V/V$, $e_{22} = e_{33} = 0$. Substitution in relation (3.8) gives

$$\tau_{11} = 2\mu\frac{\Delta V}{V} + \lambda\left(\frac{\Delta V}{V} + 0 + 0\right) = (2\mu + \lambda)\frac{\Delta V}{V}$$
$$\tau_{22} = \lambda\left(\frac{\Delta V}{V} + 0 + 0\right) = \lambda\left(\frac{\Delta V}{V}\right)$$
$$\tau_{33} = \lambda\left(\frac{\Delta V}{V} + 0 + 0\right) = \lambda\left(\frac{\Delta V}{V}\right)$$

(Note that if the stress were isotropic we would have $\tau_{11} = \tau_{22} = \tau_{33}$ as in the hydrostatic case.) $\tau_{11} - \tau_{22}$ is therefore the anisotropic part of the stress:

$$\tau_{11} - \tau_{22} = (2\mu + \lambda)\frac{\Delta V}{V} - \lambda\frac{\Delta V}{V} = 2\mu\frac{\Delta V}{V}$$

This relationship can only remain valid of course, up to the yield stress Y :

$$\boxed{\tau_{22} - \tau_{11} = -2\mu\frac{\Delta V}{V} \quad \text{for} \quad \tau_{22} - \tau_{11} \leq Y} \tag{3.9}$$

a) Assume for simplicity, that below the yield stress the stress-strain curve is linear, corresponding to constant μ.

b) For strains larger than that at the yield point, assume that the anisotropic part of the stress saturates at Y (ie; $\tau_{22} - \tau_{11} = Y$). Substituting in relation (3.9) above gives

$$Y = -2\mu\left(\frac{\Delta V}{V}\right)$$

or

$$\mu = -\frac{1}{2}\frac{Y}{\left(\frac{\Delta V}{V}\right)}$$

so that for further strain μ must decrease as the volume strain increases. Now consider the difference between the hydrostatic and x-direction pressure at the limit:

$$\tau_{11}' - \tau_{11} = \left(\frac{2}{3}\mu + \lambda\right)\frac{\Delta V}{V} - (2\mu + \lambda)\frac{\Delta V}{V} = -\frac{4}{3}\mu\left(\frac{\Delta V}{V}\right)$$

but

$$Y = -2\mu\left(\frac{\Delta V}{V}\right)$$

here, so that we can write $\tau_{11}' - \tau_{11} = 2Y/3$, (or $P_1' - P_1 = 2Y/3$) Now on the hydrostat

$$P' = -\left(\frac{2}{3}\mu+\lambda\right)\frac{\Delta V}{V}$$

but for the strain corresponding to the limit, we have

$$-\frac{\Delta V}{V} = \frac{Y}{2\mu}$$

so that

$$P_1' = \left(\frac{2}{3}\mu+\lambda\right)\frac{Y}{2\mu}$$

and the corresponding pressure on the Hugoniot is $2Y/3$ higher. $(P_1 = P_1' + 2Y/3)$ .so that

$$\boxed{P_1 = \left(\frac{2}{3}\mu+\lambda\right)\frac{Y}{2\mu}+\frac{2}{3}Y} \tag{3.10}$$

Now notice that <u>below</u> the limit, μ, λ = constant so that on the Hugoniot

$$P = -(2\mu+\lambda)\frac{\Delta V}{V}$$

and the first portion is linear. Beyond the limit however, further strain causes μ to decrease, reducing the pressure below that predicted by the linear curve. A cusp in the Hugoniot results and a two-wave structure can result. Figure 3.13 illustrates this.

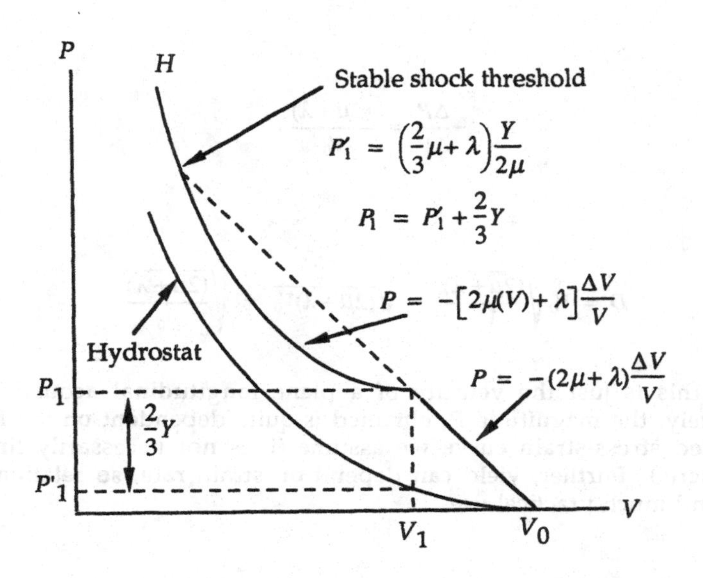

Fig. 3.13. Cusp in Hugoniot H caused by strength of materials. P_1, V_1 is the elastic limit corresponding to yield stress Y.

The first wave is called the elastic precursor and the second is the plastic wave. For driving pressures higher than that indicated for the stable shock, the elastic precursor is slower than the plastic wave, so it doesn't appear. For the elastic precursor corresponding to the state P_1, V_1 we have

$$D = V_0\sqrt{\frac{P_1 - P_0}{V_0 - V_1}} = V_0\sqrt{\frac{\Delta P}{-\Delta V}}$$

but for the one-dimensional compression

$$-\Delta P = (2\mu + \lambda)\frac{\Delta V}{V}$$

so that

$$-\frac{\Delta P}{\Delta V} = \frac{(2\mu+\lambda)}{V_0}$$

then

$$D = V_0\sqrt{\frac{(2\mu+\lambda)}{V_0}} = \sqrt{(2\mu+\lambda)V_0} = \sqrt{\frac{(2\mu+\lambda)}{\rho_0}}$$

However, this is just the velocity of a plane longitudinal acoustic wave. Unfortunately, the magnitude P_1 obtained is quite dependent on the form of the idealized stress-strain curve we assume (it is not necessarily linear as assumed here.) Further, yield can depend on strain rate, so relation (3.10) does not find much practical use.

4. Rarefaction Waves

4.1 Rarefaction Shocks

If a material has a release isentrope with a region where it reverses curvature, then there will be a region where the sound speed will be increasing as the pressure decreases. In such cases a rarefaction will steepen, causing a rarefaction shock. For normal materials however, rarefaction shocks do not exist. We will illustrate with the ideal gas. From thermodynamics we have

$$TdS \equiv \frac{\kappa_T c_v}{\beta} dP + \frac{c_P}{\beta V} dV \text{ , where } \beta = \frac{1}{V}\left(\frac{\partial V}{\partial T}\right)_P \text{, and } \kappa_T = -\frac{1}{V}\left(\frac{\partial V}{\partial P}\right)_T$$

but for the ideal gas, $\kappa_T = 1/P$, and $\beta = 1/T$ so that $\kappa_T/\beta = T/P$. We then have

$$TdS = Tc_v \ \mathrm{dln}P + Tc_P \ \mathrm{dln}V$$

or

$$dS = c_v \ \mathrm{dln}\ P + c_P \ \mathrm{dln}\ V$$

but $c_P = \gamma c_v$, so that $dS = c_v(\mathrm{dln}P + \gamma\, \mathrm{dln}V)$ or $dS = c_v \ \mathrm{dln}\ (PV^\gamma)$

and thus

$$S = c_v \ \ln\ (PV^\gamma) + \text{constant}$$

$$S - S_0 = c_v \ \ln\ (PV^\gamma) - c_v \ \ln\ (P_0 V_0^\gamma) = c_v \ \ln\left[\frac{P}{P_0}\left(\frac{V}{V_0}\right)^\gamma\right]$$

but we have already shown from relation (1.28) that

$$\frac{V}{V_0} = \frac{\mathrm{S}(\gamma-1)+(\gamma+1)}{\mathrm{S}(\gamma+1)+(\gamma-1)} = \frac{\gamma(\mathrm{S}+1)-(\mathrm{S}-1)}{\gamma(\mathrm{S}+1)+(\mathrm{S}-1)}$$

where $\mathbf{S} \equiv P/P_0$ is the shock strength. The jump in entropy across a shock in the gas is

$$S - S_0 = c_v \ln \left\{ \mathbf{S} \left[\frac{\gamma(\mathbf{S}+1) - (\mathbf{S}-1)}{\gamma(\mathbf{S}+1) + (\mathbf{S}-1)} \right]^{\gamma} \right\} \tag{4.1}$$

Note that for $\mathbf{S} \cong 1$ (very weak shocks) we have

$$S - S_0 \longrightarrow c_v \ln \left\{ 1 \left(\frac{2\gamma}{2\gamma} \right)^{\gamma} \right\} = 0$$

and we have the isentropic sound wave. Relation (4.1) shows that the entropy increases monotonically with $\mathbf{S}$ so that the stronger the shock, the greater the entropy change. Now let's assume we have a "Rarefaction shock" (ie; $\mathbf{S} < 1$). The quantity on the right of relation (4.1) becomes negative since the argument of the logarithm is less than unity. This violates the second law of thermodynamics. A rarefaction shock is thus impossible for an ideal gas.

4.2 Overtaking Release Waves

Consider a material whose Hugoniot is concave upward. We have

$$\left(\frac{\partial^2 P}{\partial V^2} \right)_H > 0 \quad \text{now} \quad c = \left(\frac{\partial P}{\partial \rho} \right)_S . \quad \text{We have} \quad c + u > D$$

Consider a shock front propagating at velocity D with a series of infinitesimal rarefactions overtaking it from the rear. Figure 4.1 shows the profile. $u_2 + c_2$ will be slightly less than $u_1 + c_1$ and so on down to the point where the particle velocity becomes zero and sound speed is c_0, because the pressure is dropping. If such a stepped profile were produced, it would soon smooth itself because the head of each step propagates faster than the foot. Until the head of the rarefaction overtakes the front of the wave, the peak pressure and the velocity at the front remain constant. After overtake, the peak pressure and velocity of the front are progressively reduced. Consider the relation between steepness of the rarefaction profile and the decay of peak pressure. We consider an approximate treatment in the interest of intuitive understanding. Figure 4.2 illustrates the relationship.

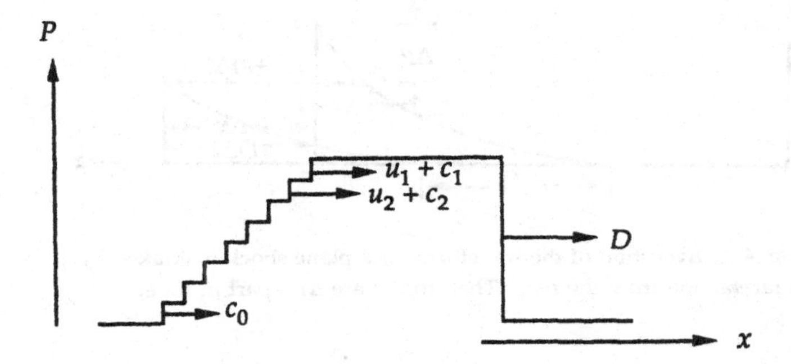

Fig. 4.1. Profile of a plane shock wave with an overtaking rarefaction. $u_1+c_1 > u_2+c_2 > 0$ so that the steps on the tail of the wave will progressively flatten. Since $u_1+c_1 > D$ the rarefaction catches the shock and will progressively weaken it.

Consider an infinitesimal time step Δt. The wave front travels a distance $\Delta x = D\Delta t$ while the point A on the back of the wave at ΔP below the peak, travels $(u+c)\Delta t$. The pressure gradient on the back of the wave is $\Delta P/\Delta s$. From the figure we have $\Delta s + \Delta x = (u+c)\Delta t$, but also $\Delta x = D\Delta t$ so that

$$\Delta s + \Delta x = (u+c)\frac{\Delta x}{D} \quad \text{or} \quad \Delta s = \left[\left(\frac{u+c}{D}\right) - 1\right]\Delta x$$

and thus

$$\frac{\Delta P}{\Delta s} = \frac{(\Delta P/\Delta x)}{\left[\left(\frac{u+c}{D}\right) - 1\right]}$$

or

$$\boxed{\frac{dP}{dx} = \left[\left(\frac{u+c}{D}\right) - 1\right]\frac{dP}{ds}} \tag{4.2}$$

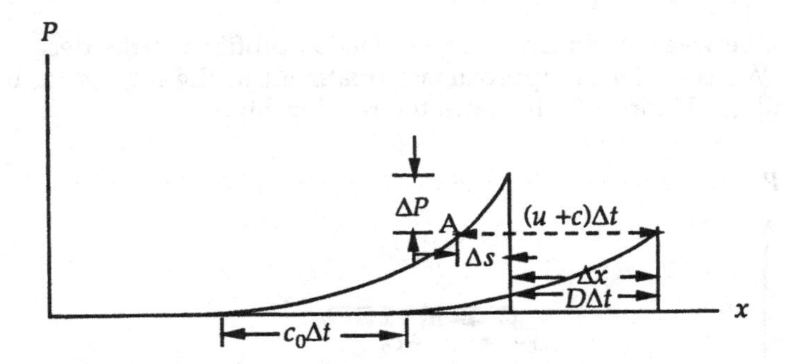

Fig. 4.2. Evolution of the waveform for a plane shock overtaken by a rarefaction from the rear. The profiles are Δt apart in time.

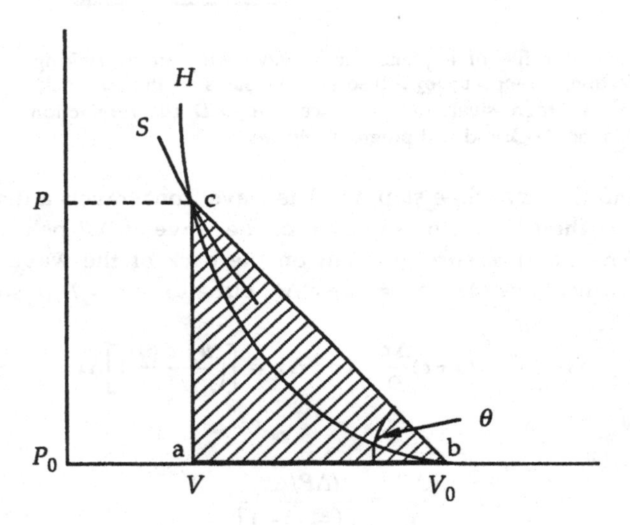

Fig. 4.3. Attenuation of a shock by a rarefaction. The velocity of the shock is given by $D = V_0 \tan^{1/2}\theta$. The sound speed is determined by the slope of the adiabat. If the Hugoniot has greater curvature, so will the adiabat, and the sound speed will thus increase.

It is thus evident that the rate of decay of the peak pressure is related to the pressure gradient on the back of the attenuating wave. Note that the attenuation here is not associated with any energy dissipation mechanism. Note also that as the wave attenuates dP/ds also decreases and so does dP/dx. Figure 4.3 illustrates the situation. Now we have

$$D = V_0\sqrt{\frac{P-P_0}{V_0-V}} = V_0 \tan^{1/2}\theta$$

$$u = \sqrt{(P-P_0)(V_0-V)} = \sqrt{2 \times \text{Area } abc}$$

and

$$c = V\left(-\frac{\partial P}{\partial V}\right)_S^{\frac{1}{2}}$$

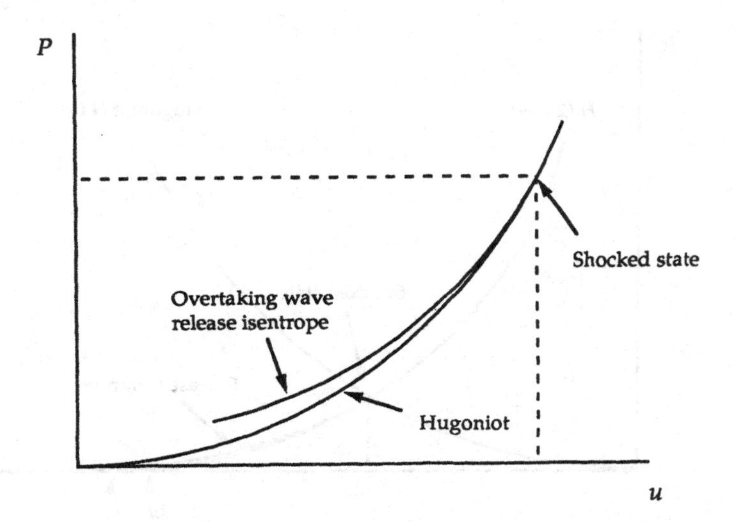

Fig. 4.4. Release by an overtaking rarefaction in the particle velocity-pressure plane.

For a given P, V both D, u are determined, while c is determined by the slope of the isentrope. If the Hugoniot has greater curvature, so will the isentrope and c will be larger. As a result $(u+c)/D$ will increase and the wave will attenuate more rapidly. In the Pressure-particle velocity plane, the release isentrope for an overtaking rarefaction wave will extend from the point on the Hugoniot corresponding to the shocked state to lower pressure and lower particle velocity as shown in figure 4.4. The pressures along the isentrope will be above the Hugoniot as a result of shock heating. The difference between the Hugoniot and the isentrope increases as the pressure decreases.

4.3 Release at a Free Surface

When release occurs at a free surface, the rarefaction is a backward going wave. The release isentrope for this case extends from the shocked state

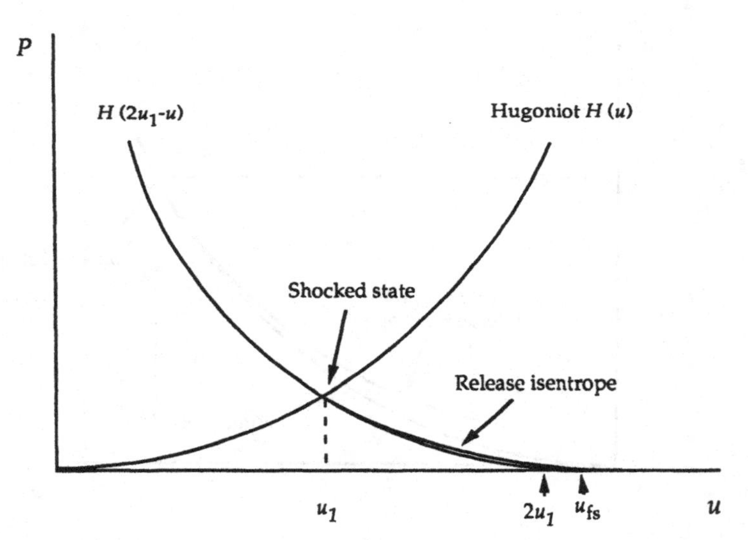

Fig. 4.5. Release at a free surface. The release isentrope is slightly above the reflection of the Hugoniot, so that the free surface velocity is slightly greater than twice the particle velocity behind the shock.

toward higher velocities. A first approximation to the release isentrope is given by the reflection of the Hugoniot in the shocked state, $H(2u_1-u)$ where u_1 is the particle velocity behind the shock. Early shock wave measurements used the measured free surface velocity as an approximation to $2u_1$. This was used to determine the particle velocity behind the shock. The true isentrope however, lies above this curve because of shock heating. Figure 4.5 shows the situation. As a numerical example[6], a common brass shocked to 1.764 MBar acquires a particle velocity of 0.269 cm/µs, and the free surface velocity is 0.564 cm/µs. The approximation gives a particle velocity of 0.282 cm/µs. This is 4.8% higher than the true value. As the shock strength increases, the approximation becomes increasingly poor.

4.4 Edge Rarefactions

If the sides of an impacted target are free, release waves will begin at the sides and move into the interior behind the shock. The shock surface, which

Table 4.1 Aluminum Data for Release from 513 kBar

P(kBar)	V_s/V_0	V_H/V_0	ρ_s(g/cm³)	ρ_H(g/cm³)
0	1.034	1.00	2.611	2.700
100	0.9250	0.9045	2.919	2.985
150	0.8886	0.8716	3.038	3.098
200	0.8578	0.8441	3.148	3.199
250	0.8319	0.8210	3.246	3.289
300	0.8092	0.8008	3.337	3.372
350	0.7892	0.7824	3.421	3.451
400	0.7710	0.7661	3.502	3.524
450	0.7543	0.7513	3.579	3.594
500	0.7392	0.7380	3.653	3.659
513	0.7340	0.7340	3.678	3.678

originally had the width of the contact surface between the target and impactor will progressively become narrower until it is finally completely

overtaken by the rarefactions. This is an important consideration in the design of experiments. Detectors intended to measure the original shock wave parameters must be located far enough from the edge so that the desired data is obtained before the rarefactions reach them. First consider the calculation of sound speed behind a shock front. We will consider aluminum shocked to 513 kBar for which McQueen, et al[7] have published a release isentrope shown in figure 4.6 and table 4.1. We take $\rho = 2.7\ \text{gm/cm}^3$. The isentrope and the Hugoniot are given by McQueen, et al as pressure versus relative volume. We will calculate the corresponding curves as a function of density to allow calculation of slope on the isentrope and the Hugoniot.
We will first fit a quadratic to each curve, then analytically calculate the slope (least squares fit):

$$P_s\,(\text{kBar}) = 712.433 - 810.713\rho + 205.921\rho^2$$

$$\left(\frac{\partial P}{\partial \rho}\right)_s = -810.713 + 411.842\rho \qquad \left(\frac{\text{kBar}\cdot\text{cm}^3}{\text{g}}\right)$$

$$P_h\,(\text{kBar}) = 1146.030 - 1120.961\rho + 258.021\rho^2$$

$$\left(\frac{\partial P}{\partial \rho}\right)_h = -1120.961 + 516.043\rho \qquad \left(\frac{\text{kBar}\cdot\text{cm}^3}{\text{g}}\right)$$

$$(2.7 \le \rho \le 3.7\ \text{g/cm}^3)$$

Some conversion of units is needed in the following form:

$$c(\text{cm}/\mu\text{s}) = \left[\left(\frac{\partial P}{\partial \rho}\right)/1000\right]^{\frac{1}{2}}$$

where the units of the derivative are $\left(\frac{\text{kBar}\cdot\text{cm}^3}{\text{g}}\right)$

We will use the densities in the previous table to calculate pressure, slope and sound speed using the fits. Results are shown in table 4.2. The corresponding results using the Hugoniot are shown in table 4.3. The last column of table

4.3 gives the approximation to the sound speed behind the shock using the Hugoniot as a substitute for the isentrope. It may be seen that the result is high by about 5%. Now let's calculate the corresponding particle velocity and shock velocity at the start of the isentrope.

$$u = u_0 \pm \sqrt{(P-P_0)(V_0-V)}\,, \text{ but } P_0 \text{ and } u_0 = 0$$

We then have

$$u^2 = P(V_0 - V) = PV_0(1 - V/V_0) = P(1 - V/V_0)/\rho_0$$

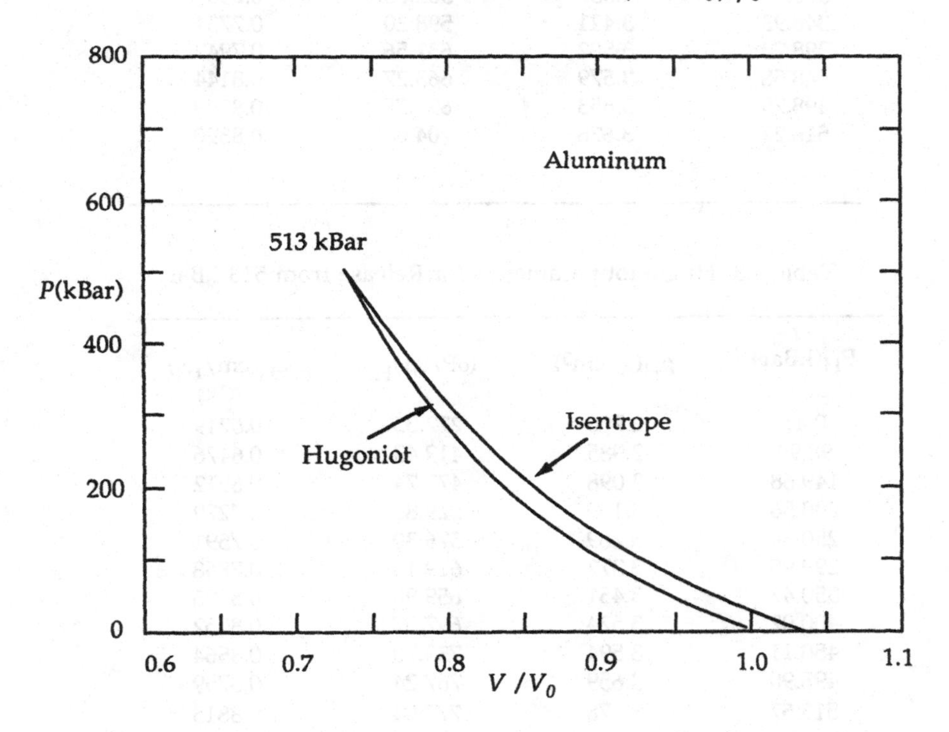

Fig. 4.6. Hugoniot and release isentrope from 513 kBar for aluminum.

Table 4.2 Isentrope Parameters for Release from 513 kBar

P_s(kBar)	ρ_s(g/cm^3)	$(\partial P/\partial \rho)_s$	c_s(cm/μs)
-0.5	2.611	264.61	0.5144
100.50	2.919	391.45	0.6257
150.00	3.038	440.46	0.6637
200.97	3.148	485.77	0.6970
250.55	3.246	526.13	0.7253
300.13	3.337	563.60	0.7507
348.93	3.421	598.20	0.7734
398.73	3.502	631.56	0.7947
448.58	3.579	663.27	0.8144
498.79	3.653	693.75	0.8329
516.27	3.678	704.04	0.8390

Table 4.3 Hugoniot parameters for Release from 513 kBar

P_H(kBar)	ρ_H(g/cm^3)	$(\partial P/\partial \rho)_H$	c_H(cm/μs)
0.41	2.700	272.35	0.5219
98.99	2.985	419.43	0.6476
149.68	3.098	477.74	0.6912
200.56	3.199	529.86	0.7279
250.34	3.289	576.30	0.7591
299.95	3.372	619.13	0.7868
350.47	3.451	659.90	0.8123
400.02	3.524	697.57	0.8352
450.11	3.594	733.70	0.8564
498.90	3.659	767.24	0.8759
513.57	3.678	777.04	0.8815

and

$$u = \sqrt{P(1-V/V_0)/\rho_0}$$

We may calculate the shock velocity:

$$D = \sqrt{\frac{P}{(1-V/V_0)\rho_0}}$$

Substituting $\rho_0 = 2.7\ \text{g/cm}^3$, and $P = 513\ \text{kBar}$, $V/V_0 = 0.734$. Corresponding to the state behind the shock, we obtain $D = 0.845\ \text{cm}/\mu\text{s}$, $u = 0.2248\ \text{cm}/\mu\text{s}$. Interpolation on the table of c_s gives $c_s = 0.838\ \text{cm}/\mu\text{s}$ at 513 kBar. This demonstrates another approximation often useful for relatively incompressible materials: $c_s \sim D$.

An alternate approximation form is also often useful. Assume $D = c_0 + su$. The sound speed is given by the Riemann invariant (to be derived later).

$$du_s = \pm\frac{1}{\rho c_s}dP_s \quad \text{so that} \quad c_s = \pm\frac{1}{\rho}\left(\frac{dP}{du}\right)_s$$

Approximate the isentrope by the Hugoniot:

$$P - P_0 = \rho_0 D u, \quad \rho_0 D = \rho(D-u)$$

$$P - P_0 \equiv \rho_0 (c_0 + su)u, \quad \frac{1}{\rho} = \frac{1}{\rho_0}\left(\frac{D-u}{D}\right)$$

$$dP = \rho_0\left[(c_0+su)du + usdu\right] = \rho_0(c_0+2su)du$$

$$\left(\frac{dP}{du}\right)_h = \rho_0(c_0+2su)$$

$$\frac{1}{\rho} = \frac{1}{\rho_0}\left[\frac{c_0+su-u}{c_0+su}\right] = \frac{1}{\rho_0}\left[\frac{c_0+(s-1)u}{c_0+su}\right]$$

Substituting gives

$$c \cong \frac{1}{\rho_0}\left[\frac{c_0+(s-1)u}{c_0+su}\right]\bullet\rho_0\,(c_0+2su)$$

or

$$\boxed{c \cong \frac{(c_0+2su)(c_0+(s-1)\,u)}{c_0+su}} \qquad (4.3)$$

Substitution of values $c_0 = 0.5386$ cm/μs, $s = 1.337$ for aluminum gives (for $u = 0.2248\ cm/\mu s$) $c \cong 0.8344$ cm/μs .

Now let's consider the use of these sound speeds for calculating edge rarefactions. We consider a plane shock wave produced by impact on a flat target by a plane flyer. The pressure at the edge of the impactor must be zero because there is no material beyond the edge to support it. We consider the progress of the wave for an elapsed time t. A rarefaction wave moves into the target from the edge and progressively weakens and retards the wave. Figure 4.7 shows the resulting geometry.

The point A with particle velocity u moves distance ut to point B, while the unretarded shock advances distance Dt . The release waves from the edge have a first arrival that advances distance ct in the material, where c is the sound speed established behind the shock. An arc of radius ct may be drawn centered on point B. The intersection of the arc with the shock at H defines angle α . The line AH is the locus of points where initial release from the shock is beginning. Release waves from deeper portions of the material, such as on line BG need not be considered, both because they are launched at later times and because they travel at slower speeds due to the attenuation of the shock front.

If we substitute

$$D = 0.845 \text{ cm}/\mu\text{s},\ u = 0.2248 \text{ cm}/\mu\text{s},\ c = 0.838 \text{ cm}/\mu\text{s}\ ,$$

from our previous results, we obtain $\alpha = 33.7^\circ$.

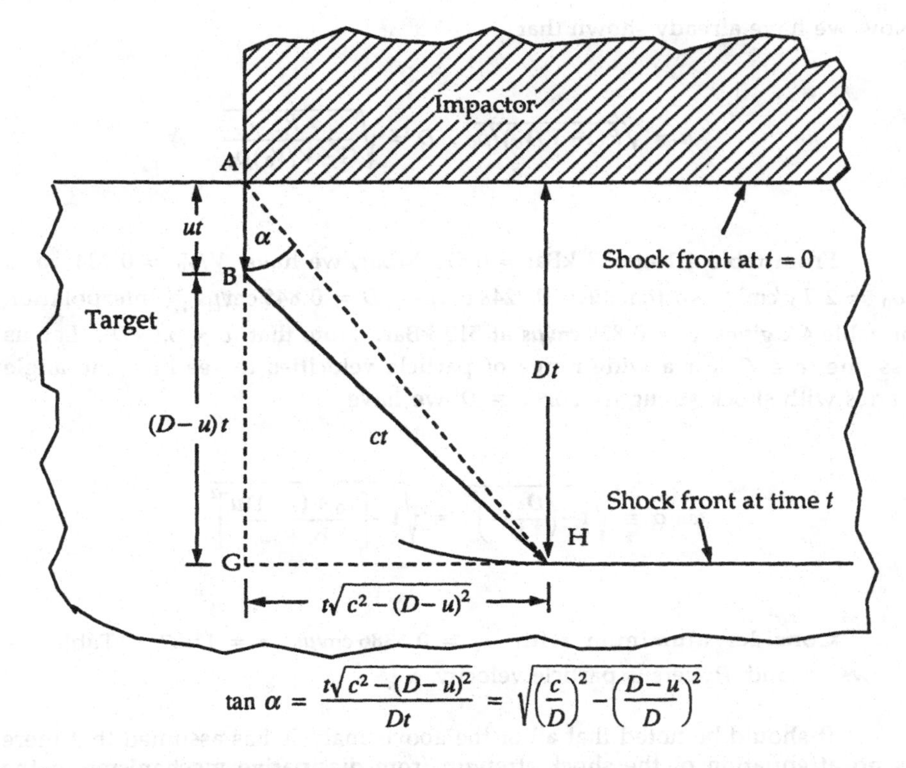

$$\tan\alpha = \frac{t\sqrt{c^2-(D-u)^2}}{Dt} = \sqrt{\left(\frac{c}{D}\right)^2-\left(\frac{D-u}{D}\right)^2}$$

Fig. 4.7. Edge rarefaction geometry. The shock front begins to acquire curvature at point H where the edge rarefaction wave has caught up with it. Point H moves inward across the shock front as time progresses. Detectors to measure shock velocity must be to the right of point H, or the measurements will not correspond to a single shock velocity.

Now we have already shown that

$$u = \sqrt{P(1-V/V_0)/\rho_0}\ ,\ D = \sqrt{\frac{P}{(1-V/V_0)\rho_0}}$$

From table 4.1 at 513 kBar = 0.513 MBar, we have $V/V_0 = 0.734$ and $\rho_0 = 2.7$ g/cm^3 , so that $u = 0.2248$ cm/μs , $D = 0.845$ cm/μs . Interpolation on table 4.2 gives $c = 0.838$ cm/μs at 513 kBar. Note that $c \approx 0.99D$. Let us assume $c \cong D$ for a wide range of particle velocities to see how the angle varies with shock strength. For $c \cong D$ we have

$$\tan\alpha \cong \sqrt{1-\left(\frac{D-u}{D}\right)^2} \cong \sqrt{1-\left[\frac{c_0+(s-1)u}{c_0+su}\right]^2}$$

Consider aluminum with $c_0 = 0.5386$ cm/μs, $s = 1.337$ Table 4.4 shows α and P versus particle velocity u .

It should be noted that all of the above analysis has assumed that there is no attenuation of the shock strength from dissipative mechanisms in the metal. Standard practice is to allow for $\alpha = 45°$. in locating electrical shorting pin detectors for measuring the steady shock transit time. Figure 4.8 shows a graphical representation for the case of aluminum shocked to 513 kBar.

If a continuously recording detector such as a velocity gauge is located in the interior of the target material, one can determine the time at which data begin to be influenced by the edge rarefactions. Figure 4.9 illustrates the analysis.

Table 4.4 angle α and shock pressure P versus particle velocity u

u (cm/μs)	α (deg)	P (MBar)
0.1	27.69	0.182
0.2	33.40	0.435
0.3	36.22	0.761
0.4	37.91	1.159
0.5	39.02	1.630
0.6	39.81	2.172
0.7	40.40	2.787
0.8	40.84	3.474
0.9	41.20	4.233
1.0	41.49	5.064

Now x_0, y_0 are the initial detector coordinates, $t_1 = y_0/D$ is the time the shock arrives at the detector, and t_2 is the time the first edge release wave arrives at the detector. Now for $t > t_1$, $y(t) = y_0 + u(t - t_1)$, and

$$y(t_2) - ut_2 = y_0 - ut_1 = y_0(1 - u/D)$$

and

$$(ct_2)^2 = x_0^2 + (y(t_2) - ut_2)^2$$

so that

$$t_2 = \frac{\sqrt{x_0^2 + y_0^2(1 - u/D)^2}}{c}$$

Note that $t = 0$ corresponds to the time the shock wave starts from the surface.

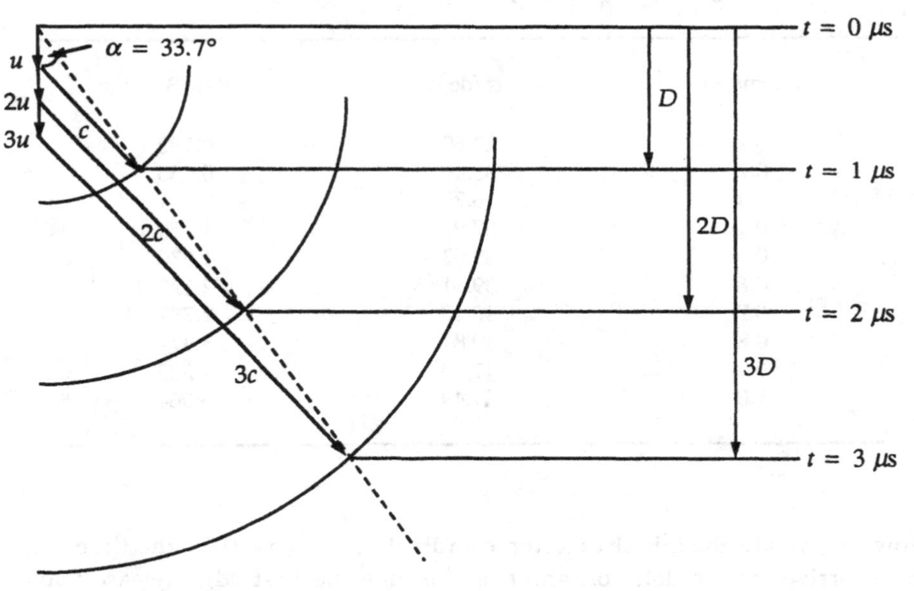

Fig. 4.8. Progression of edge release for aluminum shocked to 513 kBar. The figure is approximately to scale. The circular arcs are centered at u, $2u$, and $3u$, resp. The intersection of each arc with the corresponding plane wave gives a point on the line of first release, which is shown as a dashed curve. The distance u is 0.2685 D, determined by the ratio of the two velocities.

(Note that the analysis also assumes no overtaking release wave from the rear of the impactor has reached the detector for times $t < t_2$.)

Consider a situation where the shock wave has already been overtaken from the rear before it reaches the detector. As an example, we will consider particle velocity measurements using a Fabry-Perot velocity interferometer

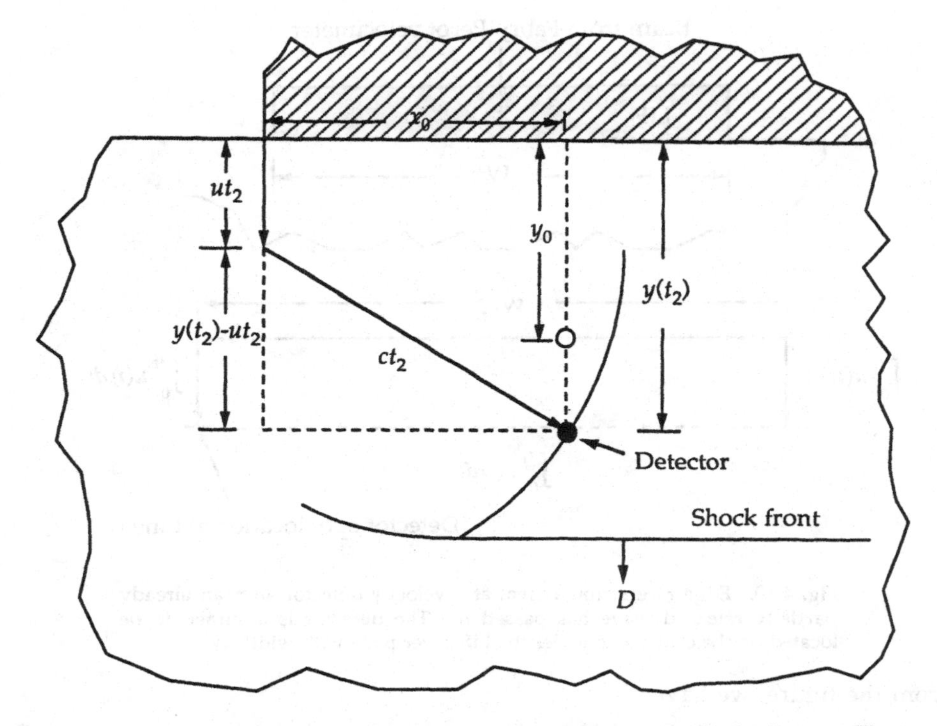

Fig. 4.9. Movement of a velocity detector during the passage of a shock wave. The initial position of the detector is at x_0, y_0. c is the sound speed behind the shock.

which measures Doppler shift of laser light returned from a moving reflective spot. We assume a lithium flouride crystal has been placed over a reflective spot on the rear of the target. We wish to determine how long it will be before edge release waves from the edge of the impactor will reach the location of the spot and influence its velocity. The principal complication is the absense of steady wave velocities. We assume t_1 to be the time the edge rarefaction reaches the spot. Since the spot moves with the material, we need only consider the horizontal distance traveled by the lead rarefaction.

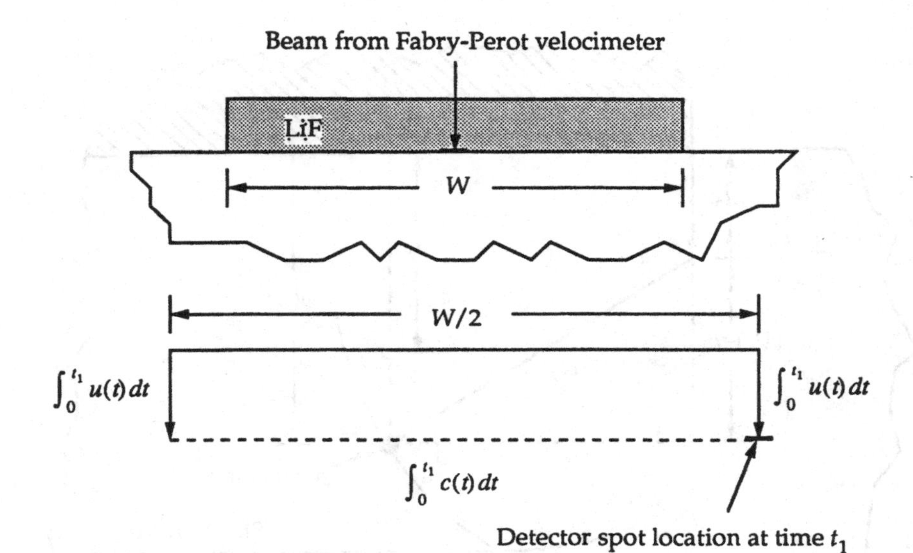

Fig. 4.10. Edge rarefaction arrival at a velocity detector after an already partially released wave has passed it. The detector is assumed to be located on the center line under the LiF cover plate with width *W*.

From the figure, we have

$$\frac{W}{2} = \int_0^{t_1} c(t)\,dt$$

where $c(t)$ is the sound velocity at a given depth in the material as the wave passes. We make the approximation $c(t) \approx D(t)$ and assume a linear D, u relation for the LiF Hugoniot (an excellent approximation.) We then have

$$\frac{W}{2} \equiv \int_0^{t_1} (c_0 + su(t))\,dt$$

We then have

$$\frac{W}{2} \cong c_0 t_1 + s\int_0^{t_1} u(t)dt$$

But we are measuring $u(t)$ with the Fabry-Perot interferometer, so that we can evaluate the integral as a running function of time either by modeling or numerical integration. We can thus find the time at which the integral equals $W/2$ to determine the approximate arrival time of the lead rarefaction from the edge of the crystal. Figure 4.11 illustrates the determination of time t_1.

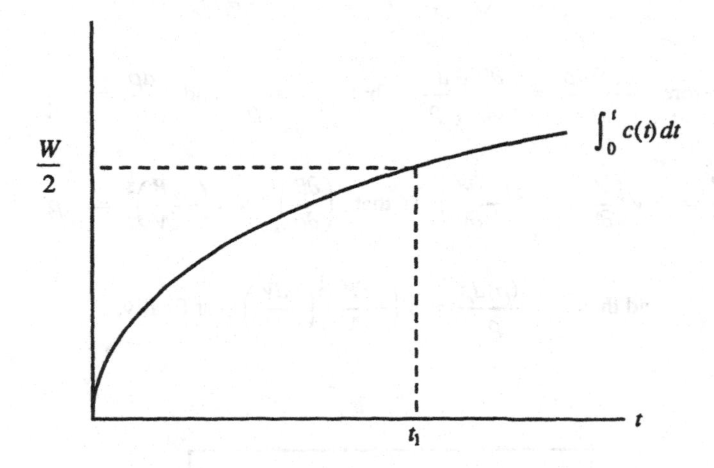

Fig. 4.11. Determination of time t_1 for which the release wave from the edge of the LiF crystal travels the distance $W/2$ to the detector.

Note that because of the impedance mismatch of the LiF crystal with the target material, the observed velocity is not that which would be seen in the bulk target material. If the target material Hugoniot is known, however, one can determine what the peak velocity in the bulk would be, and hence the corresponding pressure.

4.5 Calculation of Particle Velocity From Release Isentropes.

A number of useful forms are found in the literature for calculating the particle velocity during release if the release isentropes are known. We will derive them here. We begin with the definition of adiabatic sound speed.

$$c^2(\rho) \equiv \left(\frac{\partial P}{\partial \rho}\right)_s \quad \text{or} \quad c(\rho) = \left(\frac{\partial P}{\partial \rho}\right)_s^{\frac{1}{2}}$$

$$\text{therefore} \quad \frac{c(\rho)d\rho}{\rho} = \left(\frac{\partial P}{\partial \rho}\right)_s^{\frac{1}{2}} \frac{d\rho}{\rho} \quad \text{but} \quad V \equiv \frac{1}{\rho} \quad \text{and} \quad \frac{d\rho}{\rho} = -\frac{dV}{V}$$

$$\frac{\partial P}{\partial \rho} = -V^2 \frac{\partial P}{\partial V} = V^2\left(-\frac{\partial P}{\partial V}\right) \quad \text{so that} \quad \left(\frac{\partial P}{\partial \rho}\right)_s^{\frac{1}{2}} = V\left(-\frac{\partial P}{\partial V}\right)_s^{\frac{1}{2}} = c(\rho)$$

$$\text{and thus} \quad \frac{c(\rho)d\rho}{\rho} = V\left(-\frac{\partial P}{\partial V}\right)_s^{\frac{1}{2}}\left(-\frac{dV}{V}\right) \quad \text{or finally,}$$

$$\boxed{\frac{c(\rho)d\rho}{\rho} = -\left(-\frac{\partial P}{\partial V}\right)_s^{\frac{1}{2}} dV} \qquad (4.4)$$

Now $dP \equiv \left(\frac{\partial P}{\partial \rho}\right)_s d\rho + \left(\frac{\partial P}{\partial S}\right)_\rho dS$, but on the isentrope $dS = 0$, and $\left(\frac{\partial P}{\partial \rho}\right)_s = c^2(\rho)$, so that $dP = c^2(\rho)d\rho$ or $\frac{dP}{\rho c(\rho)} = \frac{c(\rho)d\rho}{\rho}$. But we have shown that $c(\rho) = V\left(-\frac{\partial P}{\partial V}\right)_s^{\frac{1}{2}}$ so that $\rho c(\rho) = \frac{1}{V}c(\rho) = \left(-\frac{\partial P}{\partial V}\right)_s^{\frac{1}{2}}$ and thus

$$\boxed{\frac{c(\rho)d\rho}{\rho} = \left(-\frac{\partial V}{\partial P}\right)_s^{\frac{1}{2}} dP} \tag{4.5}$$

We thus have the important result:

$$\boxed{u = \int_{\rho_0}^{\rho} \frac{c(\rho)d\rho}{\rho} = \int_{P_0}^{P} \left(-\frac{\partial V}{\partial P}\right)_s^{\frac{1}{2}} dP = -\int_{V_0}^{V} \left(-\frac{\partial P}{\partial V}\right)_s^{\frac{1}{2}} dV} \tag{4.6}$$

We have used the differential form for the Riemann invariants (to be derived in the next chapter) to relate the results to u. If the isentrope is known, we can thus calculate the particle velocity along it during the release.

4.6 Calculation of Temperature on Release Isentropes.

The TdS equations from thermodynamics give:

$$TdS \equiv c_v dT + \frac{\beta T}{\kappa_T} dV \quad \text{where} \quad \beta = \frac{1}{V}\left(\frac{\partial V}{\partial T}\right)_P \quad \text{and} \quad \kappa_T = -\frac{1}{V}\left(\frac{\partial V}{\partial P}\right)_T$$

Dividing the TdS equation by $c_v T$ gives

$$\boxed{\frac{dS}{c_v} = \frac{dT}{T} + \frac{\beta}{c_v \kappa_T} dV} \tag{4.7}$$

Along the isentrope we have $dS = 0$ so that equation 4.6 becomes

$$d \ln T = -\frac{\beta}{c_v \kappa_T} dV$$

We have

$$\ln\left(\frac{T}{T_0}\right) = -\int_{V_0}^{V} \frac{\beta}{c_v \kappa_T} dV$$

or

$$T = T_0 \exp\left(-\int_{V_0}^{V} \frac{\beta}{c_v\,\kappa_T} dV\right) = T_0 \exp\left(-\int_{V_0}^{V} \frac{\gamma_G(V)}{V} dV\right)$$

where $\gamma_G(V)$ is the Grüneisen parameter, to be derived later when we consider the Grüneisen equation of state. If we assume $\rho\gamma_G(V)$ = constant, which we will show is a good approximation for a wide range range of states (it corresponds to the ratio $\beta/c_v\,\kappa_T$), then we can write

$$\frac{\gamma_G(V)}{V} = \rho_0\,\gamma_G(V_0) \equiv \rho_0\,\gamma_0$$

and

$$\boxed{T = T_0 \exp\left(\gamma_0\left(1-\frac{V}{V_0}\right)\right)} \tag{4.8}$$

4.7 Spall

The pressure must vanish for all times at a free surface, since there is no further material to support a pressure. When a shock wave arrives at a free surface, the pressure must thus immediately drop to zero. The only way this boundary condition can be met is for a tensile wave with the same profile as the shock to be reflected back into the material. If the profile is very steep, a strong tensile stress results in the interior of the material. If this stress is greater than the strength of the material, rupture occurs, establishing a new surface. The new surface in turn must have the same boundary condition. The strength of the tensile wave continuing into the bulk of the material is reduced in the process. If the original wave was strong enough, the result may be multiple spall planes as the process repeats until the inbound tensile wave is weaker than the strength of the material. The portion of material that breaks off is called a scab. The new boundary condition at the spall plane causes a renewed pressure wave to propagate forward in the scab. When it arrives at the original free surface, the process repeats. We thus see that the scab rings as the alternate pressure and tensile waves run back and forth through it. Figure 4.12 shows the process at the initial shock wave arrival.

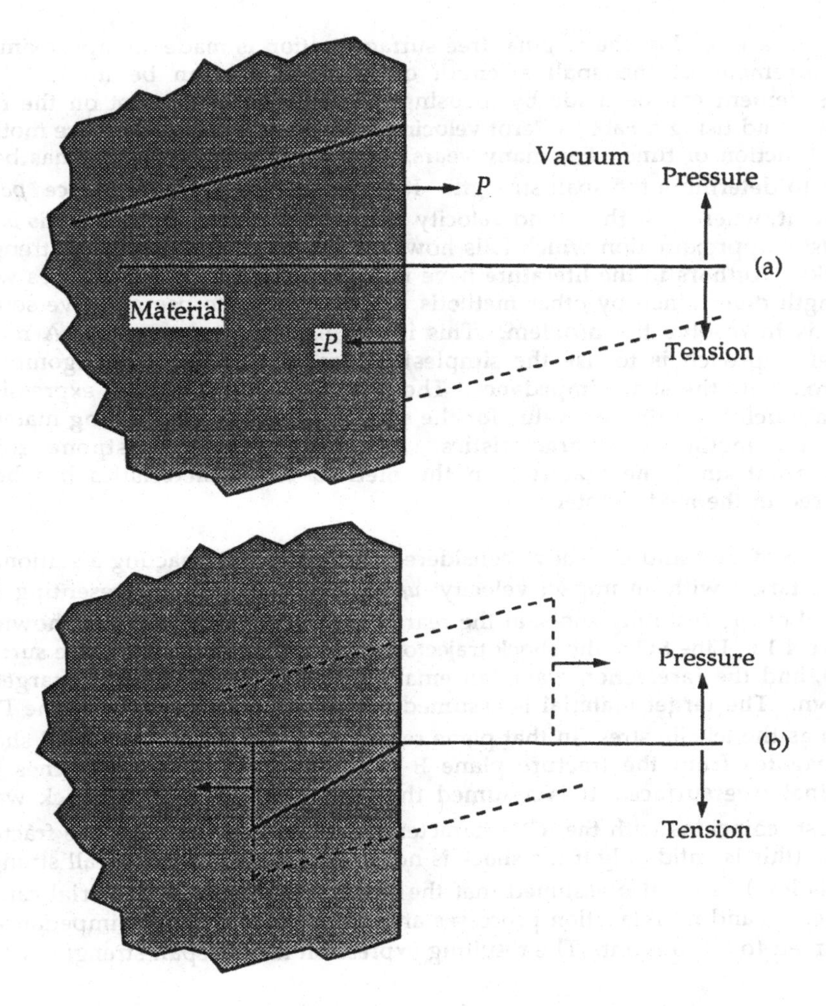

Fig. 4.12. Reflection at a free surface: (a) shows the pressure profile at the moment of emergence of the shock at the free surface, (b) shows the profile at a later time. Since the pressure must be zero at the boundary for all times, a reflected tensile wave is sent back into the material. If the tensile wave is strong enough, the material strength may be exceeded, producing a spall surface. Note that a steep tail on the back side of the shock due to overtaking rarefactions increases the peak strength in the tensile wave.

If a record of the original free surface motion is made, an approximate measurement of the spall strength of the material can be made. The measurement can be made by focusing a laser beam on a spot on the free surface and using a Fabry - Perot velocimeter to record the free surface motion as a function of time. For many years, an approximate expression has been used to determine the spall strength. It assumes the shock impedance ρc is constant, where c is the sound velocity and ρ is the mass density. This is an acoustic approximation which fails however, for even rather modest strength shocks. Authors in the literature have noted the resulting discrepancies with strength determined by other methods, and attempted to use effective sound speeds to resolve the problem. This is quite arbitrary, however. A much better approach is to use the simplest representation for the Hugoniot to approximate the shock impedance. The result is a more complex expression, but a much less arbitrary value for the spall strength. The following material uses the method of characteristics. The reader should postpone going through it until the material on the method of characteristics has been covered in the next chapter.

Novikov and Chernov[8] considered a plane flyer impacting a stationary plane target with an impact velocity u_I. The *x-t* diagram representing the arrival of the resulting shock at the rear free surface of the target is shown in figure 4.13. Line SO is the shock trajectory. The shock reaches the free surface at O, and the rarefaction wave fan emanating from O back into the target is shown. The target material is assumed to fail instantaneously at plane F as soon as the tensile stress in that plane reaches the value σ_0. The spall shock propagates from the fracture plane F within the scab until it reaches the original free surface. It is assumed that the trajectory of this shock wave almost coincides with the C^+ characteristic FK passing through the fracture plane (this is valid only if the shock is not too strong, and hence spall strength σ_0 is low). It is also assumed that the shear strength of the material can be neglected and no relaxation processes are important. The shock impedance is assumed to be constant. The resulting expression for the spall strength is

$$\sigma_0 = \frac{1}{2}\rho_0 c_0 (W_0 - W_k) \tag{4.9}$$

where ρ_0 is the initial density of the material, c_0 is the bulk sound speed, W_0 is the free-surface jump-off velocity, and W_k is the surface velocity

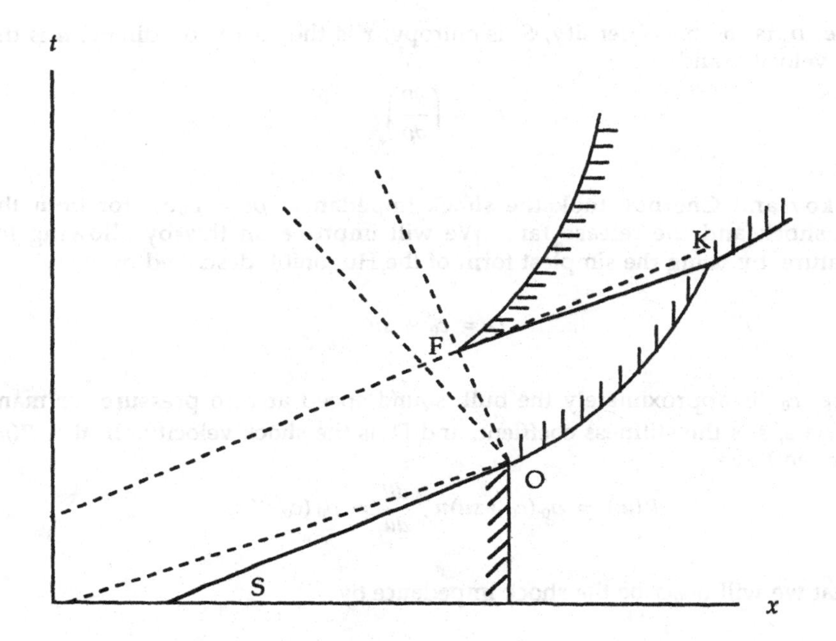

Fig. 4.13. *x-t* diagram for spall in a target impacted by a flyer. SO is the original shock trajectory. The release wave fan emanates from O. Rupture occurs at F. The spall shock then propagates to the right in the scab. Its trajectory is assumed to almost coincide with the C^+ characteristic FK passing through the fracture plane.

minimum resulting when the spall shock formed at rupture reaches the original free surface. This is the common approximate expression. Now consider development of the improved version. The equations of motion in characteristic form are:

$$
\begin{aligned}
C^+ &: \frac{dx}{dt} = u+c, \qquad dP + \rho c du = 0 \\
C^- &: \frac{dx}{dt} = u - c, \qquad dP - \rho c du = 0 \\
C^0 &: \frac{dx}{dt} = u, \qquad dS = 0
\end{aligned}
\tag{4.10}
$$

where ρ is the mass density, S is entropy, x is the space coordinate, u is the mass velocity, and

$$c^2 = \left(\frac{\partial P}{\partial \rho}\right)_s$$

Novikov and Chernov took the shock impedance $\rho c = \rho_0 c_0$ for both the spall shock and the release fan. We will improve on this by allowing for curvature, by using the simplest form of the Hugoniot, described by

$$D = c_0 + su$$

where c_0 is approximately the bulk sound speed at zero pressure for many materials, s is the stiffness coeffient, and D is the shock velocity. In the $P(u)$ plane we have

$$P(u) = \rho_0 (c_0 + su)u \, , \; \frac{dP}{du} = \rho_0 (c_0 + 2su)$$

so that we will describe the shock impedance by

$$\rho c \equiv \rho_0 (c_0 + 2su)$$

We repeat the assumption of Novikov and Chernov that the spall shock approximately follows the C^+ characteristic through the spall plane. The Riemann invariant for the C^+ characteristic can then be integrated over the path of the shock, so that

$$\int_{P_k}^{P_F} dP = -\int_{u_k}^{u_F} \rho c du$$

where F denotes the fracture plane, and k denotes the point where the spall shock reaches the free surface. Substitution of our expression for the shock impedance gives

$$P_F - P_k = -\rho_{02}\left[c_{02} + s_2 (u_F + u_k)\right](u_F - u_k) \tag{4.11}$$

where the subscript 2 indicates the target Hugoniot parameters. For the C^- characteristic extending from the first arrival of the main shock at the free surface (denoted by subscript 0) to the fracture plane, we have

$$\int_{P_0}^{P_F} dP = \int_{u_0}^{u_F} \rho c du$$

so that substitution of our approximation for the shock impedance gives

$$P_F - P_0 = \rho_{02}\left[c_{02} + s_2 (u_F + u_0)\right](u_F - u_0) \quad (4.12)$$

Subtracting relation (4.11) from relation (4.12) gives

$$P_k - P_0 = 2\rho_{02} (c_{02} + s_2 u_F) u_F - \rho_{02} c_{02} (u_0 + u_k) - \rho_{02} s_2 \left(u_0^2 + u_k^2\right) \quad (4.13)$$

Solving for u_F gives

$$u_F = -\frac{1}{2}\left(\frac{c_{02}}{s_2}\right) \pm \frac{1}{2}\left\{\left(\frac{c_{02}}{s_2}\right)^2 - 4\left[\frac{P_0 - P_k}{2\rho_{02} s_2} - \left(\frac{c_{02}}{s_2}\right)\frac{(u_0 + u_k)}{2} - \frac{(u_0^2 + u_k^2)}{2}\right]\right\}^{\frac{1}{2}}$$

Impose the boundary conditions $P_0 = P_k = 0$, $u_k = W_k$, $u_0 = W_0$, $P_F = -\sigma_0$, and take the positive branch of the radical corresponding to positive values of u_F. We obtain

$$u_F = \frac{1}{2}\left[-\left(\frac{c_{02}}{s_2}\right) + \left\{\left(\frac{c_{02}}{s_2}\right)^2 + 2\left[\left(\frac{c_{02}}{s_2}\right)(W_0 + W_k) + \left(W_0^2 + W_k^2\right)\right]\right\}^{\frac{1}{2}}\right] \quad (4.14)$$

Now from relation (4.12) with the boundary conditions applied, we have

$$\boxed{\sigma_0 = \rho_{02} (W_0 - u_F)\left[c_{02} + s_2 (W_0 + u_F)\right]} \quad (4.15)$$

Now consider the limit where s_2 approaches zero. Relations (4.11) and (4.12) become

$$P_F - P_k = -\rho_{02} c_{02} (u_F - u_k) \tag{4.16}$$

and

$$P_F - P_0 = \rho_{02} c_{02} (u_F - u_0) \tag{4.17}$$

Subtracting relation (4.16) from relation (4.17) gives

$$P_k - P_0 = \rho_{02} c_{02} (2u_F - u_0 - u_k) \tag{4.18}$$

Solving for u_F, we have

$$u_F = \frac{P_k - P_0 + \rho_{02} c_{02} (u_0 + u_k)}{2\rho_{02} c_{02}}$$

Imposing the boundary conditions gives

$$u_k = \frac{1}{2}(W_0 + W_k)$$

Substitution in relation (4.15) with $s_2 = 0$ then gives

$$\sigma_0 = \frac{1}{2}\rho_{02} c_{02} (W_0 - W_k)$$

The case $s_2 = 0$ thus corresponds to the acoustic approximation. Figure 4.14 shows the relationships in the collision and release. The initial mass velocity u_1 behind the shock is given by continuity of pressure across the shock front:

$$H_1 (u_I - u_1) = H_2 (u_1)$$

Using the linear representation for the D, u relationship to describe the Hugoniots, this becomes

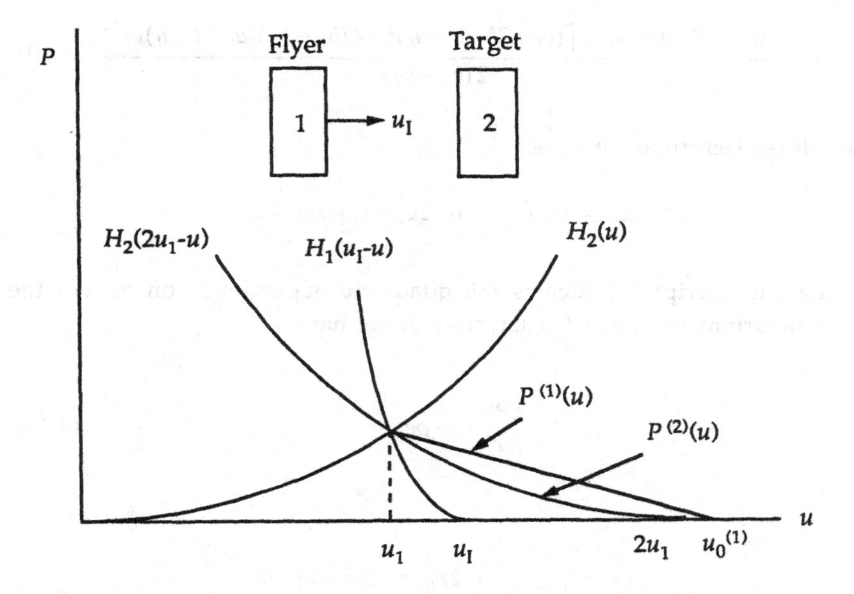

Fig. 4.14. Representation of the release isentropes. $P^{(1)}(u)$ is the acoustic approximation. $P^{(2)}(u)$ approximates the release isentrope by the target Hugoniot. $H_1(u)$ and $H_2(u)$ are the Hugoniots for the flyer and target, respectively.

$$\rho_{01}\left[c_{01} + s_1(u_I - u_1)\right](u_I - u_1) = \rho_{02}(c_{02} + s_2 u_1)u_1$$

If we simplify nomenclature by defining

$$a_1 = \rho_{01}c_{01},\ b_1 = \rho_{01}s_1,\ a_2 = \rho_{02}c_{02},\ b_2 = \rho_{02}s_2$$

and solve the resulting quadratic for u_1, we obtain

$$u_1 = \frac{(a_1 + 2b_1 u_I + a_2) - \left[(a_1 + 2b_1 u_I + a_2)^2 - 4(b_1 - b_2)(a_1 + b_1 u_I)u_I\right]^{\frac{1}{2}}}{2(b_1 - b_2)} \quad (4.19)$$

On the release isentrope we have

$$P^{(2)}(u) \equiv \rho_{02}\left[c_{02} + s_2(2u_1 - u)\right](2u_1 - u)$$

where the superscript 2 indicates the quadratic dependence on u. For the Riemann invariant on the C^+ characteristic we have

$$\frac{dP}{du} = -\rho c \quad (4.20)$$

so that

$$\rho c \equiv \rho_{02} c_{02} + 2\rho_{02} s_2 (2u_1 - u)$$

By construction, $P^{(2)}(u)$ crosses the mass velocity axis at $u = 2u_1$. For the acoustic approximation, the release isentrope assumes

$$\rho c \equiv \rho_{02} c_{02}$$

Substitution in relation (4.20) and integration gives

$$P^{(1)}(u) = -\rho_{02} c_{02} u + \text{constant}$$

But the isentrope must pass through the shocked state given by

$$P_1 = \rho_{02}(c_{02} + s_2 u_1)u_1$$

This determines the constant of integration, so that

$$P^{(1)}(u) \cong \rho_{02} c_{02} (2u_1 - u) + \rho_{02} s_2 u_1^2$$

The crossing with the mass velocity axis is given by

$$u_0^{(1)} = 2u_1 + \frac{s_2 u_1^2}{c_{02}}$$

which is somewhat above the crossing for the quadratic approximation. For symmetric collisions, where $u_1 = u_I/2$ we have

$$u_0^{(1)} = u_I + \frac{s_2}{c_{02}}\left(\frac{u_I}{2}\right)^2$$

$$P^{(1)}(u) = \rho_{02} c_{02} (u_I - u) + \rho_{02} s_2 \left(\frac{u_I}{2}\right)^2$$

$$u_0^{(2)} = 2u_1 = u_I$$

$$P^{(2)}(u) = \rho_{02}\left[c_{02} + s_2 (u_I - u)\right](u_I - u)$$

5. Method of Characteristics

5.1 Riemann Invariants

Consider the hydrodynamic flow equations

$$\frac{\partial \rho}{\partial t} + \nabla(\rho \mathbf{u}) = 0 \qquad \text{(conservation of mass)} \qquad (5.1)$$

$$\rho \frac{d\mathbf{u}}{dt} = -\nabla P \qquad \text{(Newton's second law)} \qquad (5.2)$$

where

$$\frac{d}{dt} = \frac{\partial}{\partial t} + \mathbf{u} \cdot \nabla \qquad (5.3)$$

is the derivative along a streamline of the flow. For one dimensional problems, these reduce to

$$\frac{\partial \rho}{\partial t} + \rho \frac{\partial u}{\partial x} + u \frac{\partial \rho}{\partial x} = 0 \qquad (5.4)$$

and

$$\rho\left(\frac{\partial u}{\partial t} + u \frac{\partial u}{\partial x}\right) = -\frac{\partial P}{\partial x}$$

which can be written

$$\frac{\partial u}{\partial t} + u \frac{\partial u}{\partial x} + \frac{1}{\rho}\frac{\partial P}{\partial x} = 0 \qquad (5.5)$$

now we can write

$$dP = \left(\frac{\partial P}{\partial \rho}\right)_s d\rho + \left(\frac{\partial P}{\partial S}\right)_\rho dS, \quad \text{and} \quad c^2 \equiv \left(\frac{\partial P}{\partial \rho}\right)_s$$

and for isentropic flow, we have $dS = 0$ so that $dP = c^2 d\rho$. Relation (5.4) can be written

$$\rho\frac{\partial u}{\partial x}+\frac{d\rho}{dt} = 0$$

and substituting,

$$\rho\frac{\partial u}{\partial x}+\frac{1}{c^2}\frac{dP}{dt} = 0$$

Multiply by c/ρ to obtain

$$c\frac{\partial u}{\partial x}+\frac{1}{\rho c}\frac{dP}{dt} = 0$$

or

$$c\frac{\partial u}{\partial x}+\frac{1}{\rho c}\left(\frac{\partial P}{\partial t}+u\frac{\partial P}{\partial x}\right) = 0$$

which can be written

$$\frac{1}{\rho c}\frac{\partial P}{\partial t}+\frac{u}{\rho c}\frac{\partial P}{\partial x}+c\frac{\partial u}{\partial x} = 0 \tag{5.6}$$

Add relations (5.5) and (5.6) to obtain

$$\left[\frac{\partial u}{\partial t}+(u+c)\frac{\partial u}{\partial x}\right]+\frac{1}{\rho c}\left[\frac{\partial P}{\partial t}+(u+c)\frac{\partial P}{\partial x}\right] = 0 \tag{5.7}$$

Subtract relation (5.5) from relation (5.6) to obtain

$$\left[\frac{\partial u}{\partial t}+(u-c)\frac{\partial u}{\partial x}\right]-\frac{1}{\rho c}\left[\frac{\partial P}{\partial t}+(u-c)\frac{\partial P}{\partial x}\right] = 0 \tag{5.8}$$

The form $\frac{d^+}{dt} = \frac{\partial}{\partial t} + (c+u)\frac{\partial}{\partial x}$ is the derivative along a streamline where

$$\frac{dx}{dt} = c+u$$

The form $\frac{d^-}{dt} = \frac{\partial}{\partial t} + (u-c)\frac{\partial}{\partial x}$ is the derivative along a streamline where

$$\frac{dx}{dt} = u-c$$

We can thus write relation (5.7) as

$$d^+u + \frac{1}{\rho c}d^+P = 0 \tag{5.9}$$

with

$$\frac{dx}{dt} = u+c \tag{5.10}$$

For isentropic flow we can also write $dP = c^2 d\rho$ so that

$$d^+u + \frac{c}{\rho}d^+\rho = 0 \tag{5.11}$$

Similarly,

$$d^-u - \frac{1}{\rho c}d^-P = 0 \tag{5.12}$$

$$d^-u - \frac{c}{\rho}d^-\rho = 0 \tag{5.13}$$

where

$$\frac{dx}{dt} = u - c \tag{5.14}$$

Now consider plane isentropic flow. The solutions of relation (5.10) are called C^+ characteristics and those of relation (5.14) are the C^- characteristics. For one-dimensional problems, the history of events is easily shown in an x-t diagram with a single coordinate axis and a time axis. These diagrams are highly useful for understanding the development of events and their relative timing. The x-t diagram is especially useful in understanding the method of characteristics, as will be seen in the following development. Sonic disturbances are propagated in the x-t plane along the characteristics. Intuitively, a sonic disturbance traveling at velocity c to the right relative to a medium which is itself moving to the right at velocity u, has a velocity $u + c$ in the stationary frame. Similarly, if the disturbance is traveling to the left in the same medium, its velocity is $u - c$ in the stationary frame. We thus see that the C^+ characteristics are the paths of forward sonic disturbances, while the C^- characteristics are the paths for sonic disturbances traveling in the opposite direction. Relations (5.9) and (5.11) correspond to the differential form of

$$J^+ = u + \int \frac{d^+P}{\rho c} = u + \int c \frac{d^+\rho}{\rho} \tag{5.15}$$

and relations (5.12) and (5.13) correspond to

$$J^- = u - \int \frac{d^-P}{\rho c} = u - \int c \frac{d^-\rho}{\rho} \tag{5.16}$$

J^+ and J^- are called the Riemann invariants. J^+ is constant (i.e. $dJ^+ = 0$) for flow such that

$$\frac{dx}{dt} = (u + c)$$

Now consider the case of a medium at rest, bounded at $x = 0$ by a piston which is withdrawn to the left starting at time $t = 0$. After an initial period, the piston is assumed to acquire a constant velocity. Figure 5.1

illustrates the process in the x-t plane. Along the lead C^+ characteristic, corresponding to the first sonic disturbance into the medium, we have $u = 0$, $\rho = \rho_0$ so that the lead characteristic is given by solution of

$$\frac{dx}{dt} = c(\rho_0)$$

Since it corresponds to a disturbance originating at $t = 0$, $x = 0$, this is just a straight line extending from the origin. The material to the right is undisturbed until the lead characteristic reaches it. Note that for C^- characteristics in the undisturbed material we have

$$\frac{dx}{dt} = u_0 - c(\rho_0) = -c(\rho_0) = \text{constant}$$

since the medium is initially at rest. These are thus also straight lines with negative slopes (the magnitude of the slope is the same as that of the lead C^+ characteristic), and must intersect the lead characteristic where $u = 0$ and $\rho = \rho_0$. The equation (5.13) may thus be integrated along the curve

$$\frac{dx}{dt} = (u - c)$$

to obtain u. We have

$$du^- = \frac{c}{\rho} d\rho^-, \quad \int_{u_0}^{u} du = \int_{\rho_0}^{\rho} \frac{c(\rho)}{\rho} d^-\rho$$

or since $u_0 = 0$,

$$u = \int_{\rho_0}^{\rho} \frac{c(\rho)}{\rho} d^-\rho \tag{5.17}$$

The sound velocity depends only on the density however, so this integral is independent of the path in the x-t plane. The particle velocity at a point thus depends only on the density at the point. The C^- characteristics thus all transform into a single curve $u(\rho)$ in the u, ρ dependent variable space. Such patterns of flow, in which one family of characteristics transforms to a single curve in the dependent variable plane are called simple waves. Since each point on the curve $u(\rho)$ has a corresponding $c(\rho)$ point, $u(\rho)+c(\rho)$, and thus each C^+ characteristic, corresponds to a particular density. The C^+ characteristics are thus lines of constant density. The tail characteristic is where the material attains the pressure P_1.

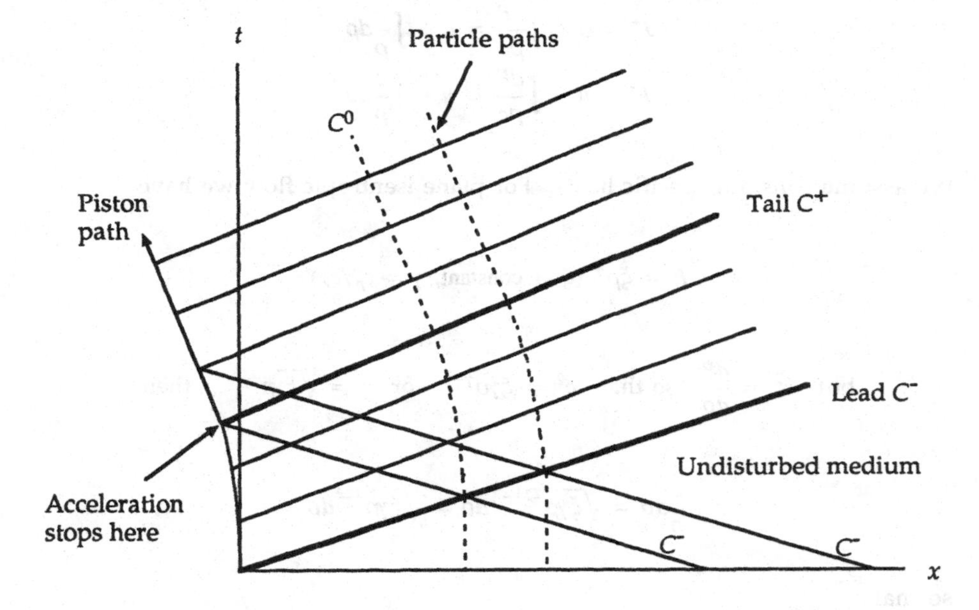

Fig. 5.1. x-t diagram illustrating the characteristics when a piston is smoothly withdrawn to a final velocity behind a gas. Regions in the gas to the right of the lead C^+ characteristic have not yet seen the disturbance. Regions beyond the tail C^+ characteristic have reached a final velocity. The dotted paths are the trajectories of the mass points.

$$\frac{d^+x}{dt} = u+c = -\int_{\rho_0}^{\rho_1} \frac{c(\rho)}{\rho} d^+\rho + \left(\frac{\partial P}{\partial \rho}\right)_{s,\, P=P_1}^{\frac{1}{2}}$$

For normal materials, u and c both decrease as density decreases, so that ρ or P decrease monotonically with $d^{+}x/dt$. A mass point is undisturbed until its position is reached through a monotonically decreasing pressure gradient between the lead and tail C^{+} characteristics. The region between the lead and tail characteristics is called a rarefaction fan.

Now let's consider a practical illustration using the ideal gas. The Riemann invariants are

$$J^{+} = u + \int \frac{dP}{\rho c} = u + \int \frac{c}{\rho} d\rho$$

$$J^{-} = u - \int \frac{dP}{\rho c} = u - \int \frac{c}{\rho} d\rho$$

We assume constant specific heats. For plane isentropic flow we have

$$P = \xi \rho^{\gamma} \;\; (\xi = \text{constant}, \; \gamma = c_p / c_v)$$

but $c^2 = \dfrac{dP}{d\rho}$ so that $c^2 = \xi \gamma \rho^{\gamma - 1}$ or $c = \sqrt{\xi \gamma}\, \rho^{\frac{\gamma-1}{2}}$, then

$$\frac{c}{\rho} d\rho = \sqrt{\xi \gamma}\, \rho^{\frac{\gamma-1}{2} - 1} d\rho = \sqrt{\xi \gamma}\, \rho^{\frac{\gamma-3}{2}} d\rho$$

so that

$$\int \frac{c}{\rho} d\rho = \sqrt{\xi \gamma} \frac{2}{(\gamma - 1)} \rho^{\frac{\gamma-1}{2}} = \frac{2}{(\gamma - 1)} c$$

and thus

$$\boxed{J^{\pm} = u \pm \frac{2}{(\gamma - 1)} c} \qquad (5.18)$$

We can describe the flow either in terms of u and c or J^+ and J^-. We have

$$J^+ = u + \frac{2}{(\gamma-1)}c$$

$$J^- = u - \frac{2}{(\gamma-1)}c$$

so that

$$u = \frac{1}{2}(J^+ + J^-)$$

and since

$$J^+ - J^- = \frac{4}{(\gamma-1)}c$$

we have

$$\boxed{\begin{aligned} u &= \frac{1}{2}(J^+ + J^-) \\ c &= \frac{(\gamma-1)}{4}(J^+ - J^-) \end{aligned}} \tag{5.19}$$

Now

$$u+c = \left(\frac{1}{2}+\frac{\gamma-1}{4}\right)J^+ + \left(\frac{1}{2}-\frac{\gamma-1}{4}\right)J^- = \left(\frac{1+\gamma}{4}\right)J^+ + \left(\frac{3-\gamma}{4}\right)J^-$$

and

$$u-c = \left(\frac{1}{2}-\frac{\gamma-1}{4}\right)J^+ + \left(\frac{1}{2}+\frac{\gamma-1}{4}\right)J^- = \left(\frac{3-\gamma}{4}\right)J^+ + \left(\frac{1+\gamma}{4}\right)J^-$$

We can write

$$\frac{dx}{dt} = u+c = F^+(J^+, J^-) \quad \text{on } C^+$$

and

$$\frac{dx}{dt} = u - c = F^-(J^+, J^-) \quad \text{on } C^-$$

where

$$F^+ = \left(\frac{\gamma+1}{4}\right)J^+ + \left(\frac{3-\gamma}{4}\right)J^-$$
$$F^- = \left(\frac{3-\gamma}{4}\right)J^+ + \left(\frac{\gamma+1}{4}\right)J^-$$

In general, the functions F^+ and F^- are determined by the properties of the medium. Now on C^+ we have J^+ = const., so that the slope dx/dt for the C^+ characteristic is determined only by J^-. Similarly, on the C^- characteristic the slope is determined by J^+.

Now consider plane isentropic flow in an infinite space. Assume that at $t = 0$ the distribution of flow variables u, c along the x axis are given (initial conditions). We have $u(x, 0), c(x,0)$. We see that this determines $J^+(x, 0), J^-(x, 0)$. The flow variables u and c at point D in figure 5.2 are seen to be given by

$$u(\mathrm{D}) = \frac{1}{2}\left[J^+(\mathrm{D})+J^-(\mathrm{D})\right]$$
$$c(\mathrm{D}) = \frac{(\gamma-1)}{4}\left[J^+(\mathrm{D})-J^-(\mathrm{D})\right]$$

but $J^+(\mathrm{D}) = J^+(\mathrm{A}), J^-(\mathrm{D}) = J^-(\mathrm{B})$. Note that the position of D depends on the paths of the C^+ and C^- characteristics from A and B so that we cannot say conditions at D are determined only by those at A and B. The slope at G is determined by the C^- characteristic from point E. The conditions at D are thus determined by the initial conditions on segment AB. If we are considering continuous flow then u and c must be unique at point D. Since these are uniquely related to J^+, J^- there, we can have only one C^+ and one C^- characteristic at any point D(x,t). This means that the characteristics of a given family (+ or -) do not cross anywhere for such flow. Under these conditions, the situation at D is determined exclusively by the conditions of the segment AB. (In situations where characteristics of a family cross each other, the values of J^+ and J^- are not unique, hence neither are u and c. We thus have discontinuous conditions (i.e. shock waves).

5.2 Region of Influence

Figure 5.3 shows the full region of influence of the segment AB. The C^- characteristic from A and the C^+ characteristic from B bound the region of influence of the segment AB for continuous flow. A point such as G cannot

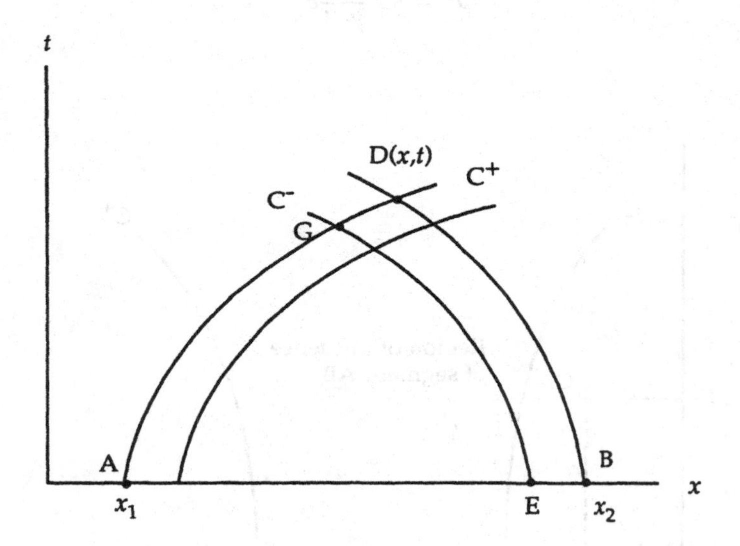

Fig. 5.2. x-t diagram for continuous isentropic flow. Conditions at the point D are determined entirely by the initial conditions on the segment AB.

be influenced by the initial conditions on segment AB since disturbances from AB cannot reach $x = x_G$ until later times. Now consider the previous figure. We have

$$J^+(\mathrm{D}) = J^+(\mathrm{A}) = J^+(x_1,0)$$
$$J^-(\mathrm{D}) = J^-(\mathrm{B}) = J^-(x_2,0)$$

But for the ideal gas with constant specific heats we have shown that

$$J^{\pm} = u \pm \frac{2}{\gamma - 1} c$$

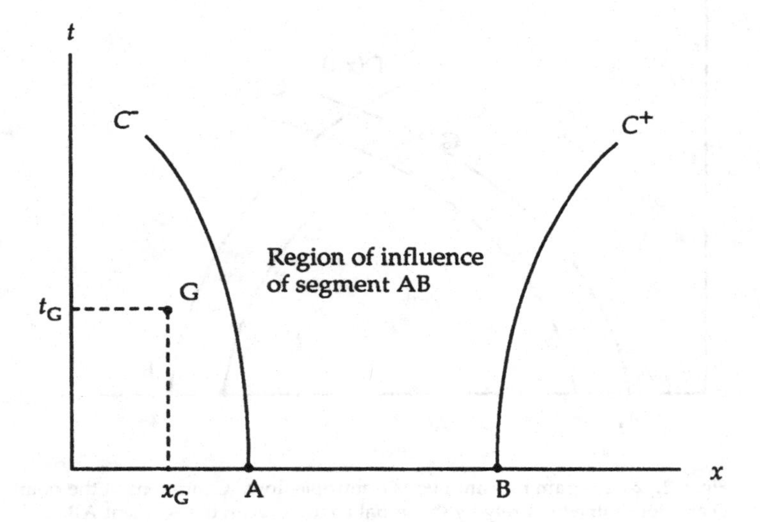

Fig. 5.3. *x-t* diagram for continuous flow. The conditions on the segment AB influence the entire region between the C⁺ characteristic emanating from point B and the C⁻ characteristic emanating trom point A. The point G is entirely unaffected by conditions on segment AB.

so that

$$J^+(x_1,0) = u_1 + \frac{2}{\gamma-1}c_1, \text{ and } J^-(x_2,0) = u_2 - \frac{2}{\gamma-1}c_2$$

and finally,

$$\boxed{\begin{aligned} u(x,t) &= \left(\frac{u_1+u_2}{2}\right)+\frac{2}{\gamma-1}\left(\frac{c_1-c_2}{2}\right) \\ c(x,t) &= \left(\frac{c_1+c_2}{2}\right)+\frac{\gamma-1}{2}\left(\frac{u_1-u_2}{2}\right) \end{aligned}} \tag{5.20}$$

Now $P = \xi\rho^\gamma$ and $c^2 = \xi\gamma p^{\gamma-1}$ so that

$$\left(\frac{c^2}{\xi\gamma}\right)^{\frac{1}{\gamma-1}} = \rho \quad \text{or} \quad \rho = \left(\frac{1}{\xi\gamma}\right)^{\frac{1}{\gamma-1}} c^{\frac{2}{\gamma-1}}$$

and thus

$$\xi\rho^\gamma = \xi\left(\frac{1}{\xi\gamma}\right)^{\frac{\gamma}{\gamma-1}} c^{\frac{2\gamma}{\gamma-1}}$$

The pressure at the point x,t is thus also determined.

$$\boxed{P(x,t) = \left[\frac{\xi}{(\xi\gamma)^{\frac{\gamma}{\gamma-1}}}\right] c(x,t)^{\frac{2\gamma}{\gamma-1}}} \tag{5.21}$$

In general, for an arbitrary point D(x,t), the conditions at D are determined by those along an arbitrary curve AB shown in figure 5.4, cutting the C^+ and C^- characteristics passing through the point.

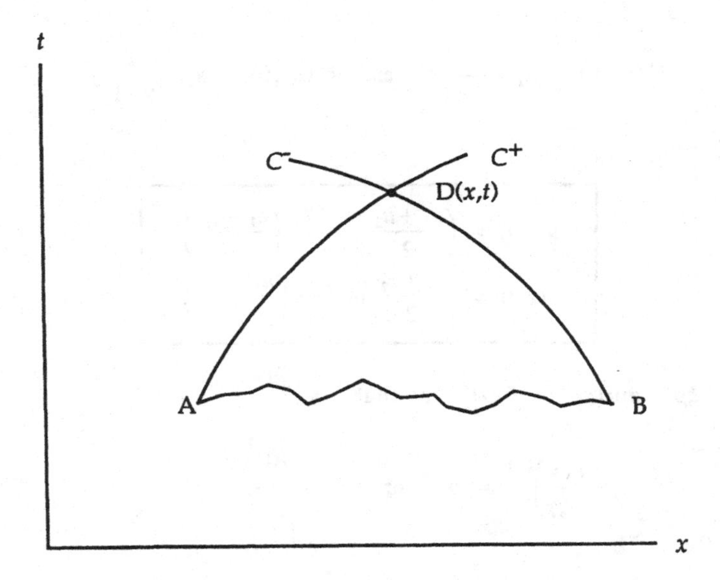

Fig. 5.4. Conditions at point D are determined by those along an arbitrary curve AB cutting the C^+ and C^- characteristics passing through the point.

5.3 Analytic Example Using the Ideal Gas

Consider an ideal gas with constant specific heats bounded on the left by a piston. The gas is initially at rest. At time zero the piston is withdrawn with constant acceleration until time t_B , after which the piston moves at constant velocity. See figure 5.5 for the analysis. (x,t) is an arbitrary field point in the figure. (x^*,t^*) is an arbitrary point on the piston track. (x_B,t_B) is the point where the acceleration ends at point B. v_B is the piston velocity at point B. Now we have

$$J^{\pm} = u \pm \frac{2}{\gamma-1}c$$

But for the undisturbed gas, we have $u_0 = 0$, $c = c_0$, therefore,

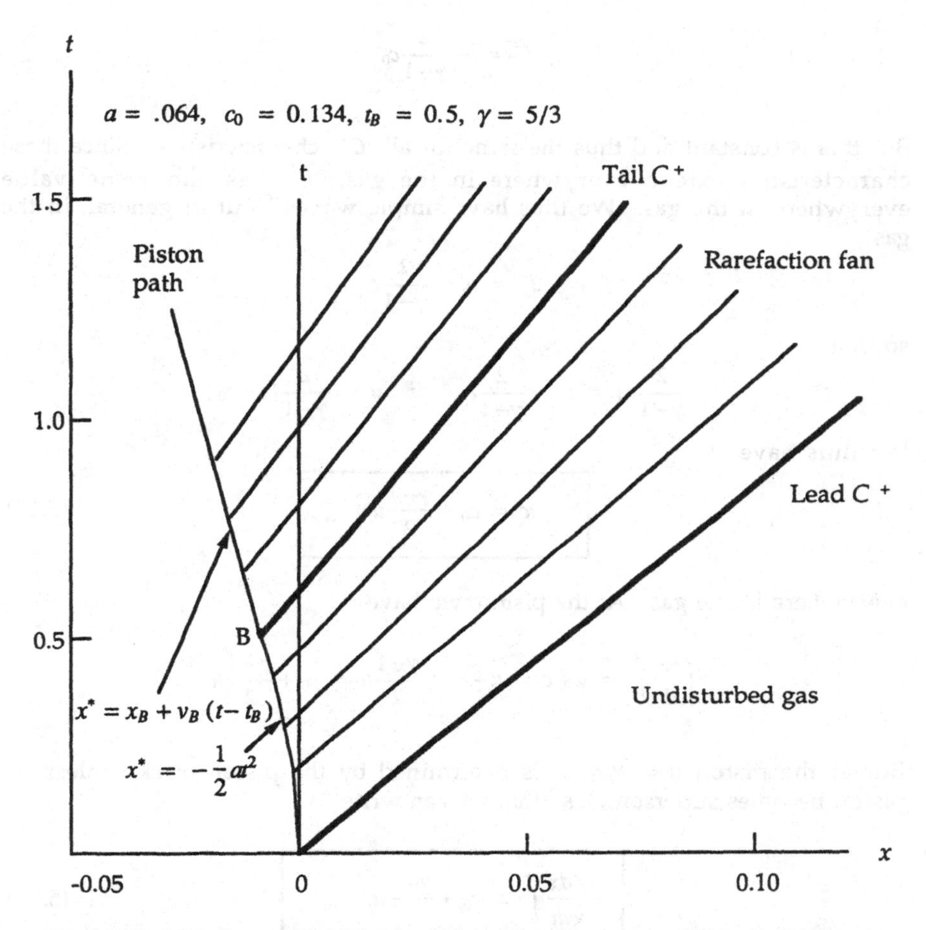

Fig. 5.5. Numerical example of a piston being withdrawn from an ideal gas with constant acceleration a from time 0 to time 0.5, and constant velocity thereafter. The assumed acceleration and the sound speed and specific heat ratio of the gas are shown in the figure. The figure is only roughly drawn to scale.

$$J^- = -\frac{2}{\gamma-1}c_0$$

But this is constant and thus the same for all C^- characteristics. Since these characteristics extend everywhere in the gas, J^- has the same value everywhere in the gas. We thus have simple waves. But in general, in the gas

$$J^- = u - \frac{2}{\gamma-1}c$$

so that

$$-\frac{2}{\gamma-1}c_0 = u - \frac{2}{\gamma-1}c \quad \text{or} \quad u = \frac{2}{\gamma-1}(c-c_0)$$

We thus have

$$\boxed{c = c_0 + \frac{\gamma-1}{2}u} \tag{5.22}$$

everywhere in the gas. At the piston, we have

$$\left(\frac{dx}{dt}\right)^+ = u+c = u+c_0+\frac{\gamma-1}{2}u = c_0+\frac{\gamma+1}{2}u$$

But at the piston $u = u_{piston}$ is determined by the piston track (unless the piston becomes supersonic), so that we can write

$$\boxed{\left(\frac{dx}{dt}\right)^+ = c_0+\frac{\gamma+1}{2}u_{piston}} \tag{5.23}$$

everywhere on the track. Now consider the accelerated portion where

$$x_B^* = -\frac{1}{2}at^{*2}, \quad v_B = -at^* = u_{piston}$$

$$\left(\frac{dx}{dt}\right)^{+}_{t<t_B} = c_0 - \frac{\gamma+1}{2}at^*$$

$$x_B = -\frac{1}{2}st_B^2, \; v_B = -at_B$$

The C^+ characteristics are straight lines extending from the track at position (x^*, t^*) with a slope that depends on the position on the track. The characteristic is described by

$$x = x^* + \left(\frac{dx}{dt}\right)^{+}_{t<t_B}(t-t^*)$$

or since $x^* = -\frac{1}{2}at^{*2}$ and $\left(\frac{dx}{dt}\right)^{+}_{t<t_B} = c_0 - \frac{\gamma+1}{2}at^*$ the characteristic is described by

$$\boxed{x = \left[c_0 - \frac{\gamma+1}{2}at^*\right](t-t^*) - \frac{1}{2}at^{*2}} \tag{5.24}$$

For the coasting segment of the track $(t > t_B)$ the track is given by

$$x^* = x_B + v_B(t-t_B)$$

The C^+ characteristics are straight lines, all with the same slope, extending from the track at (x^*, t^*) :

$$x = x^* + \left(\frac{dx}{dt}\right)^{+}_{t>t_B}(t-t^*)$$

but $x^* = x_B + v_B(t-t_B)$ so that

$$\boxed{x = x_B + v_B(t-t_B) + \left[c_0 + \frac{\gamma+1}{2}v_B\right](t-t^*)} \tag{5.25}$$

It should be noted that <u>all</u> the C^+ characteristics are straight lines because J^+ is constant on a given C^+ characteristic and J^- is a constant <u>everywhere</u>, hence both u and c are constant on the characteristic. As a result, $(dx/dt)^+$ is constant for any given C^+ characteristic. We have shown that

$$c = c_0 + \frac{\gamma - 1}{2} u$$

everywhere in the gas, and

$$\left(\frac{dx}{dt}\right)^- \equiv u - c$$

so that

$$\left(\frac{dx}{dt}\right)^- = u - c_0 + \frac{1-\gamma}{2} u = \left[1 + \frac{1}{2} - \frac{\gamma}{2}\right] u - c_0 = \frac{3-\gamma}{2} u - c_0$$

and

$$\boxed{\left(\frac{dx}{dt}\right)^- = \left(\frac{3-\gamma}{2}\right) u - c_0} \tag{5.26}$$

everywhere in the gas. Now since the C^+ characteristics are straight lines, for a C^+ originating at x^*, t^*, we have

$$\left(\frac{dx}{dt}\right)^+ = c_0 + \frac{\gamma+1}{2} u = c_0 + \frac{\gamma+1}{2} u_B = c_0 - \frac{\gamma+1}{2} a t^*$$

as the slope <u>anywhere</u> on that characteristic. Consequently we have $u = -at^*$ <u>anywhere</u> on the characteristic emanating from x^*, t^*. The C^+ characteristics are thus <u>iso-velocity</u> curves. We have

$$\left(\frac{dx}{dt}\right)^0 = -at^*, \quad \left(\frac{dx}{dt}\right)^- = -c_0 - \left(\frac{3-\gamma}{2}\right) a t^*, \quad \left(\frac{dx}{dt}\right)^+ = c_0 - \left(\frac{\gamma+1}{2}\right) a t^*$$

everywhere on the C^+ characteristic from x^*, t^* in the fan. For the lead characteristic where $u = 0$, we have $t^* = 0$ and

$$\left(\frac{dx}{dt}\right)^{-}_{lead} = -c_0, \quad \left(\frac{dx}{dt}\right)^{0}_{lead} = 0, \quad \left(\frac{dx}{dt}\right)^{+}_{lead} = +c_0$$

For the tail characteristic, where $t^* = t_B$ we have

$$\left(\frac{dx}{dt}\right)^{-}_{tail} = -c_0 - \left(\frac{3-\gamma}{2}\right)at_B, \quad \left(\frac{dx}{dt}\right)^{0}_{tail} = -at_B, \quad \left(\frac{dx}{dt}\right)^{+}_{tail} = c_0 - \left(\frac{\gamma+1}{2}\right)at_B$$

Now $v_B = -at_B$ so that we can write

$$\left(\frac{dx}{dt}\right)^{-}_{tail} = -c_0 + \left(\frac{3-\gamma}{2}\right)v_B, \quad \left(\frac{dx}{dt}\right)^{0}_{tail} = v_B, \quad \left(\frac{dx}{dt}\right)^{+}_{tail} = c_0 + \left(\frac{\gamma+1}{2}\right)v_B$$

and these values are maintained for all subsequent C^+ characteristics. All points beyond the tail C^+ characteristic are moving at the final piston velocity. (The derivative with the superscript zero refers to the J^0 characteristics which we have not discussed yet. These represent the particle paths.) Figure 5.5 shows a numerical example using arbitrary units chosen merely to give a convenient graphical representation. We assume a monatomic ideal gas ($\gamma = 5/3$), and assume acceleration $a = .064$, sound speed $c_0 = 0.134$, and coast time $t_B = 0.5$.

5.4 Approximate Methods

For short distances in the x-t plane, the characteristics can be approximated by straight lines. If we take two points on the x axis sufficiently close together, we can calculate the initial slopes of the characteristics and project the corresponding straight lines to an intersection at a point D. Figure 5.6 illustrates the calculation.

For point D we solve simultaneously for the time t :

$$(u_1 + c_1)t_D + x_1 = (u_2 - c_2)t_D + x_2$$

We have

$$t_D = \frac{x_2 - x_1}{u_1 + c_1 - u_2 + c_2}, \quad x_D = (u_1 + c_1)t_D + x_1$$

If we assume an ideal gas with constant specific heats, we have

$$u_D = \left(\frac{u_1 + u_2}{2}\right) + \frac{2}{\gamma - 1}\left(\frac{c_1 - c_2}{2}\right), \quad c_D = \left(\frac{c_1 + c_2}{2}\right) + \frac{\gamma - 1}{2}\left(\frac{u_1 - u_2}{2}\right)$$

Once a line of points such as D has been calculated for adjacent points x_1, x_2, x_3,... the analysis can be iterated to extend new characteristics from the points such as D. The next line of points will be designated D*. Figure 5.7 illustrates the calculation of the next line of points. We have

$$(u_{D_1} + c_{D_1})(t^* - t_{D_1}) + x_{D_1} = (u_{D_2} - c_{D_2})(t^* - t_{D_2}) + x_{D_2}$$

or

$$(u_{D_1} + c_{D_1} - u_{D_2} + c_{D_2})t^* = (u_{D_1} + c_{D_1})t_{D_1} - (u_{D_2} - c_{D_2})t_{D_2} + x_{D_2} - x_{D_1}$$

and thus

$$t^* = \frac{(u_{D_1} + c_{D_1})t_{D_1} - (u_{D_2} - c_{D_2})t_{D_2} + x_{D_2} - x_{D_1}}{(u_{D_1} + c_{D_1} - u_{D_2} + c_{D_2})}$$

$$x^* = (u_{D_1} + c_{D_1})(t^* - t_{D_1}) + x_{D_1}$$

$$u_D^* = \left(\frac{u_{D_1} + u_{D_2}}{2}\right) + \frac{2}{\gamma - 1}\left(\frac{c_{D_1} - c_{D_2}}{2}\right)$$

$$c_D^* = \left(\frac{c_{D_1} + c_{D_2}}{2}\right) + \frac{\gamma - 1}{2}\left(\frac{u_{D_1} - u_{D_2}}{2}\right)$$

Now consider the gas to be bounded by surfaces which, in general, will be allowed to move. We assume however, that the gas remains in contact (ie; the walls are not moving at supersonic speed.) The track of the bounding surfaces is given by functions $\phi_1(t)$ and $\phi_2(t)$. The gas velocity at the bounding surfaces is just that of the walls, which is determined by $\phi_1(t)$ and $\phi_2(t)$. Figure 5.8 illustrates the calculation.

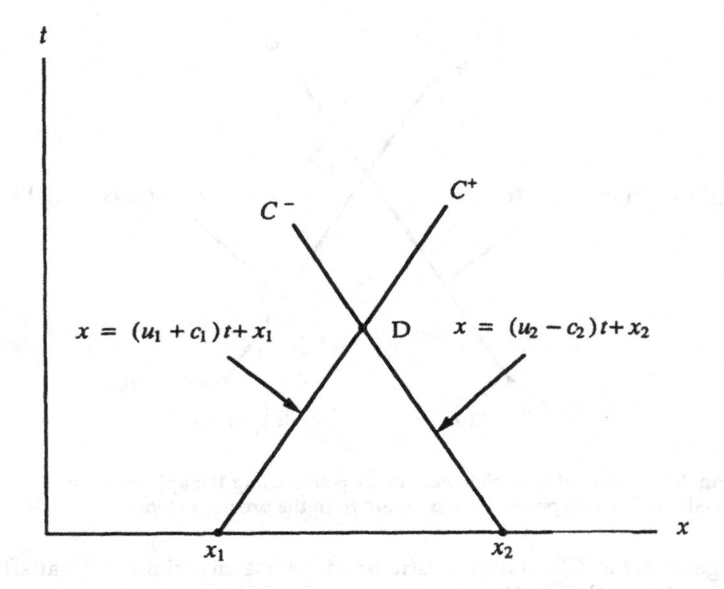

Fig. 5.6. Approximate method of calculationg a subsequent point in the *x-t* plane assuming constant sound speeds and particle velocities for short intervals. The two points x_1 and x_2 give initial conditions at two adjacent points. The approximate location of point D is calculated by linearly extrapolating the characteristics from the two points until they intersect.

At the point M we have

$$J^-(\mathrm{M}) = J^-(\mathrm{A}) \quad \text{and} \quad u_\mathrm{M} = \phi_1'(t) \quad \text{where} \quad \phi' \equiv d\phi/dt$$

We thus have the boundary conditions. Now

$$u_\mathrm{M} = \frac{1}{2}(J^+(\mathrm{M}) + J^-(\mathrm{M})) \quad \text{or with} \quad u_\mathrm{M} = \phi_1'(t)$$

we have

$$\boxed{J^+(\mathrm{M}) = 2\phi_1'(t) - J^-(\mathrm{A})} \tag{5.27}$$

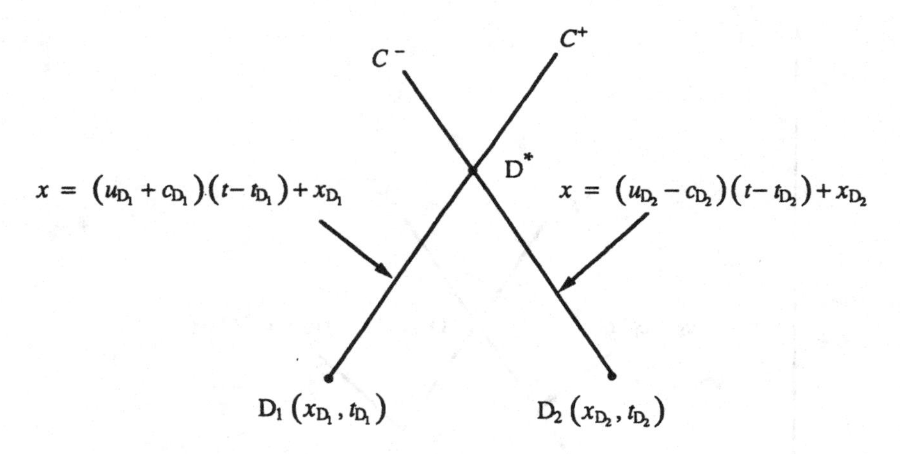

Fig. 5.7. Calculation of the next line of points using the approximate method. The two points D_1 and D_2 are from the previous step.

We generate a C^+ characteristic at M whose invariant J^+ satisfies the boundary condition there. We also have

$$c_M = \frac{\gamma - 1}{4}(J^+(M) - J^-(M))$$

x_M, t_M are given by the intersection of the C^- characteristic from A with the track $x = \phi_1(t)$. We can approximate the C^+ characteristic from M by

$$x = (u_M + c_M)(t - t_M) + x_M$$

and find the intersection with the C^- characteristic from point B, to find point D. The C^+ characteristic can be extended further for successive points in the same manner. For the right side boundary at F we have

$$u_F = \phi_2'(t), \quad J^+(F) = J^+(G)$$

Since

$$u_F = \frac{1}{2}(J^+(F) + J^-(F))$$

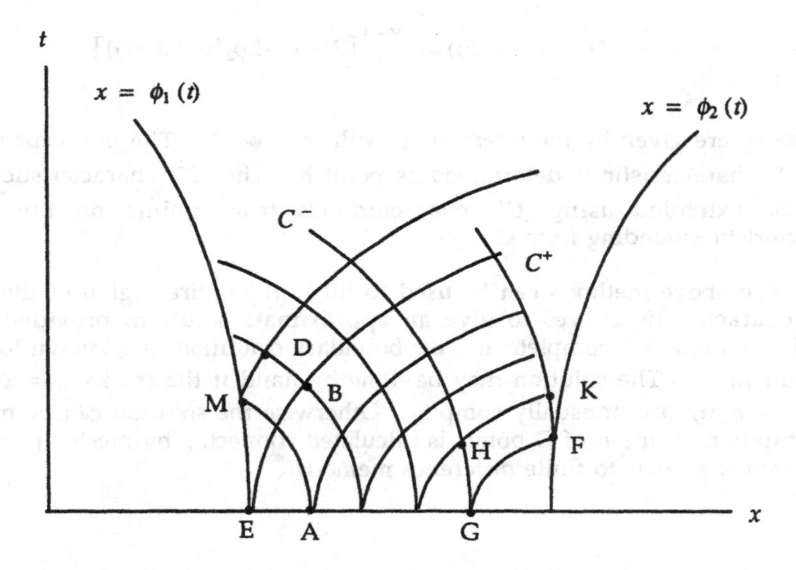

Fig. 5.8. Influence of the bounding surfaces on the characteristics for later times in the approximate method. Surfaces $\phi_1(t)$ and $\phi_2(t)$ are the tracks of the bounding surfaces of the gas. It is assumed that the motion of the bounding surfaces is subsonic, so that the gas remains in contact with them.

we have

$$J^-(\text{F}) = 2u_\text{F} - J^+(\text{F})$$

or

$$\boxed{J^-(\text{F}) = 2\phi_2{}'(t) - J^+(\text{G})} \tag{5.28}$$

We thus generate a C^- characteristic from F whose J^- invariant satisfies the boundary condition there. The C^- characteristic from F can be approximated by

$$x = (u_\text{F} - c_\text{F})(t - t_\text{F}) + x_\text{F}$$

where

$$c_F = \frac{\gamma-1}{4}(J^+(F)-J^-(F)) = \frac{\gamma-1}{4}\left[J^+(G)-2\phi_2'(t)+J^+(G)\right]$$

and x_F, t_F are given by the intersection with $x = \phi_2(t)$. The intercept with the C^+ characteristic is determined as point K. The C^- characteristic can then be extended using C^+ characteristics from points on the C^- characteristic extending from G.

The above methods can be used to fill in the entire region of the gas with characteristic curves to give an approximate solution, provided the initial conditions are complete and the boundary conditions are known for all relevant times. The solution may be done by hand if the tracks $x = \phi_1(t)$ and $x = \phi_2(t)$ are unusually complex. Otherwise the solution can be made by computer. A mesh of D points is calculated connected by mesh equations in a manner similar to finite difference methods.

5.5 Simple Waves

Consider a gas bounded on the left side by a piston which is withdrawn along track $x_1 = \psi_1(t)$. At initial time $t = 0$, let $J^-(x,0) =$ constant in the entire region occupied by the gas $(x > x_{10})$. figure 5.9 illustrates the x-t diagram. At subsequent times J^- will also remain constant in the entire region bounded by the piston, $x > x_1 = \psi_1(t)$. (The piston only excites C^+ characteristics; the C^- characteristics arrive at the boundary from the past. The piston sends only J^+ invariants into the future.)

The values of J^- in the entire x-t plane occupied by the gas are determined by the initial values of J^- on the x axis. (If the piston were on the right, $x_2 = \psi_2(t)$, $x_{20} = \psi_2(0)$, and $J^+(x,0) =$ constant for $x < x_{20}$, then J^+ = constant for $x < x_2 = \psi_2(t)$.) We have shown that

$$\left(\frac{dx}{dt}\right)^{C^+} = F^+(J^+,J^-), \quad \left(\frac{dx}{dt}\right)^{C^-} = F^-(J^+,J^-)$$

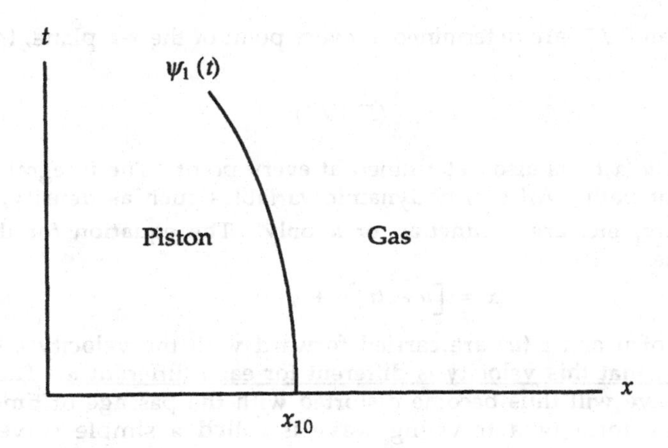

Fig. 5.9. ***x-t*** **diagram for a gas bounded on the left by a piston which is withdrawn at subsonic speed along the track** $\psi_1(t)$**.** $J^+(x,0)$ **= constant in the entire region occupied by the gas** ($x < x_{10}$)**. This produces simple waves.**

On the C^+ characteristic we have J^+ = constant, but for our case we also have J^- = constant so that the slope

$$\left(\frac{dx}{dt}\right)^{C^+} = \text{constant}$$

The C^+ characteristics are thus straight lines of the form

$$x = F^+(J^+, J^-)t + \phi(J^+)$$

where ϕ is the constant of integration associated with the particular characteristic. It is determined by the initial and boundary conditions. Now

$$J^- = u - \int \frac{dP}{\rho c} = \text{constant}$$

or

$$u = \text{constant} + \int \frac{dP}{\rho c}$$

But since J^+ and J^- are determined at every point of the x-t plane, from

$$u = \frac{1}{2}(J^+ + J^-)$$

it follows that $u\,(x,t\,)$ is also determined at every point. The integral is thus independent of path. All thermodynamic variables such as density, sound speed, pressure, etc. are a function of u only. The equation for the C^+ characteristic is

$$x = \left[u + c(u)\right]t + \phi(u)$$

Given values of u and $c\,(u)$ are carried forward with the velocity $u + c\,(u)$. Note however, that this velocity is different for each different u. The initial profile of a wave will thus become distorted with the passage of time. The solution in the form of a traveling wave is called a simple wave. The solution is a wave traveling to the right, defined by

$$u = f\left[x - (u + c(u))t\right], \quad c = g\left[x - (u + c(u))t\right]$$

where the forms of f and g are determined by the initial and boundary conditions. The initial profiles of $u\,(x,\,0)$, $c\,(x,\,0)$ become distorted with the passage of time. For a simple wave traveling to the left, we have J^+ = constant throughout the gas and J^- = constant on the C^- characteristics so that the C^- characteristics are straight lines of the form

$$x = F^-(J^+,\, J^-)t + \psi(J^-)$$

We have

$$J^+ = u + \int \frac{dP}{\rho c} = \text{constant}$$

so that the integral is again independent of path. We have

$$x = (u - c(u))t + \psi(u)$$

or

$$u = f'\left[x + (c(u) - u)t\right], \quad c = g'\left[x + (c(u) - u)t\right]$$

6. Hugoniots for Porous Materials

Consider a three term equation of state (EOS). We will use the Grüneisen model for the thermal pressure contribution. This will be treated later in more detail in a separate section.

$$E = E_c(V) + E_T + E_e, \quad P = P_c(V) + P_T + P_e$$

$$E_c(V) = \int_V^{V_{0c}} P_c(V)dV, \quad E_T = 3Nk(T - T_0) + E_0$$

$$E_e = \frac{1}{2}\beta_0\left(\frac{V}{V_0}\right)^{\frac{1}{2}} T^2, \quad P_e = \frac{1}{2}\frac{E_e}{V}$$

$$P_T = \frac{\gamma_G(V)E_T}{V} \quad \text{(Gruneisen model)}$$

β_0 = electronic specific heat at standard volume
E_0 = thermal energy of the lattice at room temperature
T_0 = room temperature
$E_c(V)$ = elastic energy from cold compression curve
E_e = electronic energy contribution
E_T = thermal energy of the lattice at temperature T
$P_c(V)$ = cold compression curve (0°K)
P_e = electronic pressure contribution
P_T = thermal pressure

Now consider the case where we assume $P_0 \sim 0$ and ignore electronic contributions. (This limits the upper pressures for which analysis will be valid.) From the Hugoniot equations, we have

$$E - E_0 = \frac{1}{2}(P + P_0)(V_0 - V) \cong \frac{1}{2}P(V_0 - V)$$

We will also neglect E_0 and P_0

$$E \cong E_c + E_T \ , \quad P \cong P_c + P_T = P_c + \gamma_G(V)\frac{E_T}{V}$$

Then

$$(P - P_c)\frac{V}{\gamma_G(V)} = E_T$$

so that

$$E \cong E_c + \frac{V(P - P_c)}{\gamma_G(V)}$$

Substituting in the Hugoniot equations and neglecting E_0 gives

$$E_c + \frac{V(P - P_c)}{\gamma_G(V)} \cong \frac{1}{2}P(V_0 - V)$$

We now wish to solve for the pressure P. First multiply by $2/V$:

$$\frac{2E_c}{V} + \frac{2P}{\gamma_G(V)} - \frac{2P_c}{\gamma_G(V)} = P\left(\frac{V_0}{V} - 1\right)$$

or

$$P\left[\left(\frac{V_0}{V} - 1\right) - \frac{2}{\gamma_G(V)}\right] = \frac{2E_c}{V} - \frac{2P_c}{\gamma_G(V)}$$

$$P\left[\left(\frac{2}{\gamma_G(V)} + 1\right) - \frac{V_0}{V}\right] = \frac{2P_c}{\gamma_G(V)} - \frac{2E_c}{V}$$

then

$$P = \frac{\left(\frac{2}{\gamma_G(V)}\right)P_c - \frac{2E_c}{V}}{\left(\frac{2}{\gamma_G(V)} + 1\right) - \frac{V_0}{V}}$$

$$\text{Define} \quad \xi \equiv \frac{2}{\gamma_G(V)} + 1 \quad \text{so that} \quad \frac{2}{\gamma_G(V)} = \xi - 1$$

We have

$$\boxed{P = \frac{(\xi(V) - 1) P_c(V) - 2E_c(V)/V}{\xi(V) - V_0/V}} \tag{6.1}$$

Now $\rho \gamma_G(V) \equiv$ constant (commonly observed behavior) so that

$$\frac{\gamma_G(V)}{V} \equiv \frac{\gamma_0}{V_0}, \qquad \gamma_G(V) \equiv \gamma_0 \left(\frac{V}{V_0}\right)$$

then

$$\frac{2}{\gamma_G(V)} = \frac{V_0}{V}\left(\frac{2}{\gamma_0}\right) \quad \text{and} \quad \xi = \frac{V_0}{V}\left(\frac{2}{\gamma_0}\right) + 1 \quad \text{so that}$$

$$\xi(V) - \frac{V_0}{V} = \frac{V_0}{V}\left(\frac{2}{\gamma_0}\right) - \frac{V_0}{V} + 1 = \frac{V_0}{V}\left(\frac{2}{\gamma_0} - 1\right) + 1$$

The limiting compression is seen to occur when

$$\xi(V) - \frac{V_0}{V} = 0 \quad \text{or} \quad \left(\frac{V_0}{V}\right)_{\text{Lim}} \left(\frac{2}{\gamma_0} - 1\right) + 1 = 0$$

We have

$$\left(\frac{V_0}{V}\right)_{\text{Lim}} = \left(\frac{\rho}{\rho_0}\right)_{\text{Lim}} = \frac{-1}{\left(\frac{2}{\gamma_0} - 1\right)} = \frac{1}{1 - \frac{2}{\gamma_0}} = \frac{\gamma_0/2}{\gamma_0/2 - 1} = \frac{\gamma_0}{\gamma_0 - 2}$$

$$\boxed{\left(\frac{\rho}{\rho_0}\right)_{\text{Lim}} = \frac{\gamma_0}{\gamma_0 - 2}} \tag{6.2}$$

The Hugoniot takes the form

$$P = \frac{\frac{V_0}{V}\left(\frac{2}{\gamma_0}\right)P_c(V) - 2E_c(V)/V}{\frac{V_0}{V}\left(\frac{2}{\gamma_0} - 1\right) + \frac{V}{V}}$$

or

$$\boxed{P = \frac{2V_0 P_c(V)/\gamma_0 - 2E_c(V)}{V + V_0(2/\gamma_0 - 1)}} \tag{6.3}$$

Note that none of the assumptions above assumed the material was continuous. For porous material, we can take the initial specific volume as $V_{00} > V_0$ so that

$$\boxed{P_H(V, V_{00}) = \frac{2V_{00} P_c(V)/\gamma_0 - 2E_c(V)}{V + V_{00}(2/\gamma_0 - 1)}} \tag{6.4}$$

We thus have a <u>family</u> of Hugoniots depending on the initial density (or porosity). For no porosity $(V_{00} = V_0)$, we have the principal Hugoniot of the continuous material. Figure 6.1 illustrates the family in the P, V plane. We have $P_c(V_0) = 0$, $E_c(V_0) = 0$ so that $P_H(V_0) = 0$. At $V = V_0$ we have

$$P_H(V_0, V_{00}) = \frac{2V_{00} P_c(V_0)/\gamma_0 - 2E_c(V_0)}{V_0 + V_{00}(2/\gamma_0 - 1)} = 0$$

since the numerator vanishes. Thus all members of the family emanate from the same point $P = 0$, $V = V_0$ in the P, V plane. In shock compression of porous materials, the initial step is to crush the pores, heating the material until it becomes continuous, then jumping to the final state. (If slow compression were performed the crushing would occur at low pressures until V_0 was reached, followed by compression on an isentrope.) Also, at low pressures, material strength may prevent complete crush-up of the pores. The pressure asymptote for each member of the family is given by

$$V_{Lim}^{(i)} = \omega_i V_0 \left(1 - \frac{2}{\gamma_0}\right)$$

where ω is the porosity. For highly porous material, the thermal pressures produced in crushing the pores with shock waves can become so great as to result in anomalous Hugoniots: the final density can be less than the initial density of the <u>continuous</u> material (ie; we can have $\rho_{00} < \rho < \rho_0$.) It should be noted that relation (6.4) applies only to the initial behavior of the Hugoniots. At large pressures the electronic contributions cannot be ignored.

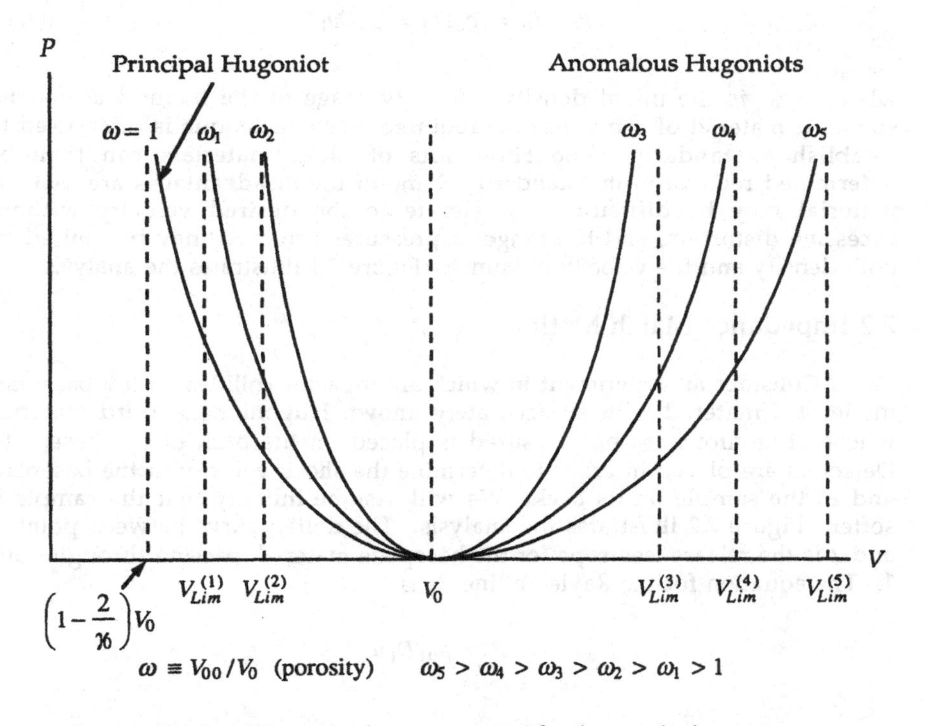

Fig. 6.1. Family of Hugoniots for porous materials. As porosity increases, we obtain anomalous Hugoniots which give final densities greater than that of the initial density of the continuous material as a result of heating produced in the crushing pores.

7. Hugoniot Measurement Methods

7.1 Direct Collision Method

In this method, the flyer and the target are both of the material to be measured. It is also called the symmetric method. One measures the velocity of the shock wave in the target in lab frame (D) and the impactor velocity u_I. By symmetry, the particle velocity u will be $u_I/2$ (conservation of momentum.) A series of measurements defines the Hugoniot in either the P, u or D, u planes. Conservation of momentum gives

$$P - P_0 = \rho_0 D u = \rho_0 D u_I / 2 \tag{7.1}$$

where ρ_0 is the initial density. An advantage of the method is that no standard material of comparison is required. This technique is often used to establish a standard. The Hugoniots of other materials can then be determined relative to the standard. Some of the disadvantages are that the material may be difficult to accelerate to the desired velocity without excessive distortion, and the range of pressures achieved may be limited by both density and the velocity of launch. Figure 7.1 illustrates the analysis.

7.2 Impedance Match Method

Consider an experiment in which an impactor collides with a baseplate made of a material with an accurately known Hugoniot. A third material, whose Hugoniot is to be measured is placed on the back of the baseplate. Detectors are placed in order to determine the shock velocity in the baseplate and in the sample on its back. We will assume initially that the sample is softer. Figure 7.2 illustrates the analysis. The dotted curve between points 1 and 2 is the release isentrope for the baseplate material, passing through point 1. The equation for the Rayleigh line R_1 is

$$P = \rho_{01} D_1 u$$

where D_1 is the measured shock velocity in the baseplate and ρ_{01} is its initial density. The equation for the Rayleigh line R_2 is

$$P = \rho_{02} D_2 u$$

where D_2 is the measured shock velocity in the sample, and ρ_{02} is the initial density of the sample. The intersection of R_2 with the isentrope from

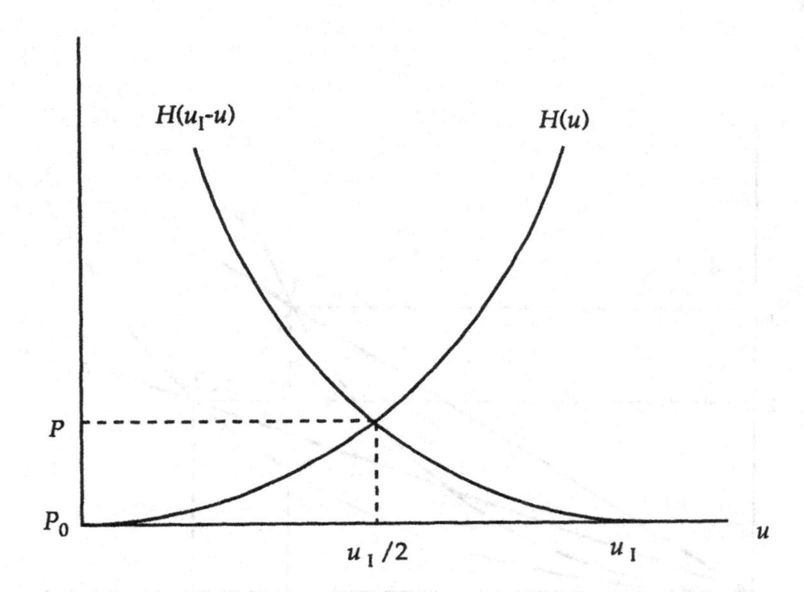

Fig. 7.1. Determination of the shocked state produced in a symmetric collision of two identical materials. The target is assumed to be at rest, while the impactor has velocity u_I. Conservation of momentum gives the particle velocity behind the shock front as half of the initial velocity of the impactor.

point 1 gives point 2 on the principal Hugoniot of the sample. Now consider the case where the sample is a stiffer material than the baseplate (i.e. it has a stiffer Hugoniot.) Figure 7.3 illustrates the analysis for this case. The analysis is as before, but the curve 1 - 2 is on the reshock Hugoniot of the baseplate material ($P_2 > P_1$). Thus far, for all normal materials, the reshock Hugoniot observed has fallen on the reflection of the principal Hugoniot in the velocity u_1 of the state established by the first shock.

7.3 Reshock Method

The impactor velocity may be measured by either x-ray pictures in flight for sufficiently opaque materials, or perhaps collision with a stepped target visible to a streak camera. Figure 7.4 shows the geometrical arrangement.

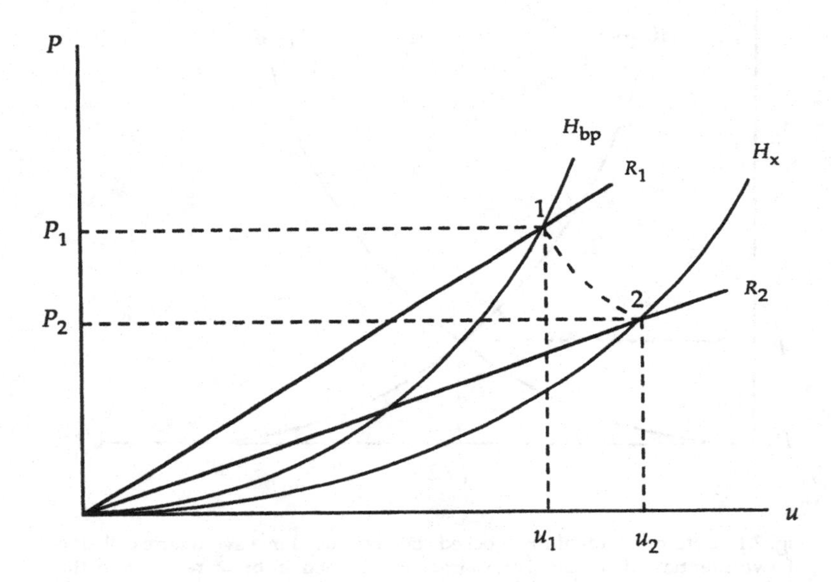

Fig. 7.2. Impedance match measurements with a baseplate that has a stiffer Hugoniot than that of the sample. H_x is the Hugoniot of the sample. H_{bp} is the Hugoniot of the baseplate. R_1 and R_2 are the Rayleigh lines.

Figure 7.5 illustrates the analysis. The Rayleigh line R_3 is given by

$$P = \rho_{03} D u$$

Measurement of u_I, with the known Hugoniots $H_1(u)$, $H_2(u)$ establishes point 1 describing the state produced by the first shock. The measurement of D establishes the Rayleigh line R_3, and the intersection with the known Hugoniot $H_3(u)$ establishes point 2 on the reshock Hugoniot H_2' of material 2. Thus far, all measurements of normal materials have been consistent with

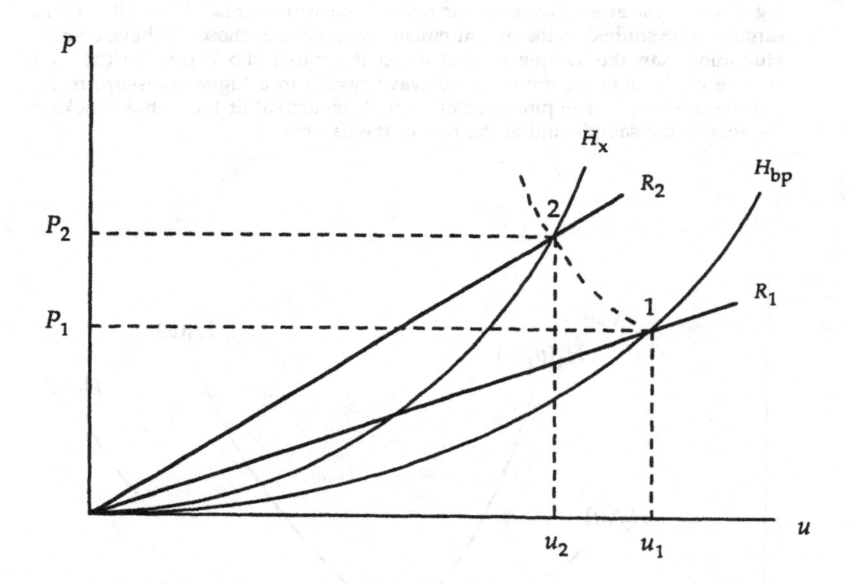

Fig. 7.3. Impedance match measurements with a baseplate that has a softer Hugoniot than that of the sample. H_x is the Hugoniot of the sample. H_{bp} is the Hugoniot of the baseplate. R_1 and R_2 are the Rayleigh lines.

$$H_2'(u) = H_2(2u_1 - u)$$

so that one can treat point 3 as approximately another point on $H_2(u)$.

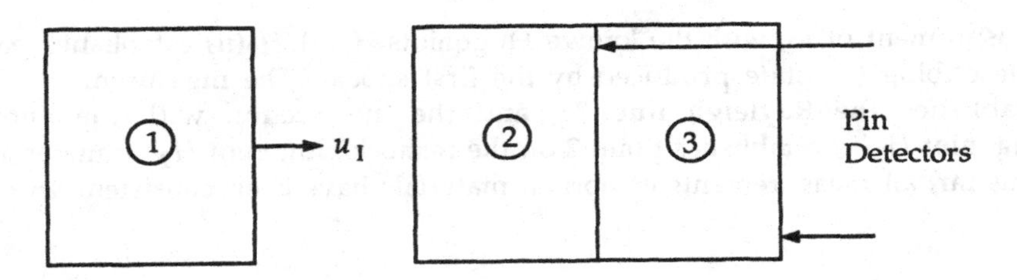

Fig. 7.4. Physical arrangement for reshock measurements. Material 2 is the sample to be studied in the measurement. Material 3 is chosen to have a stiffer Hugoniot than the sample so that when the initial shock reaches the back surface of the sample, the reflected wave results in a higher pressure in the sample (reshock). The pin detectors signal the arrival of the initial shocks at the rear of the sample and at the rear of the baseplate.

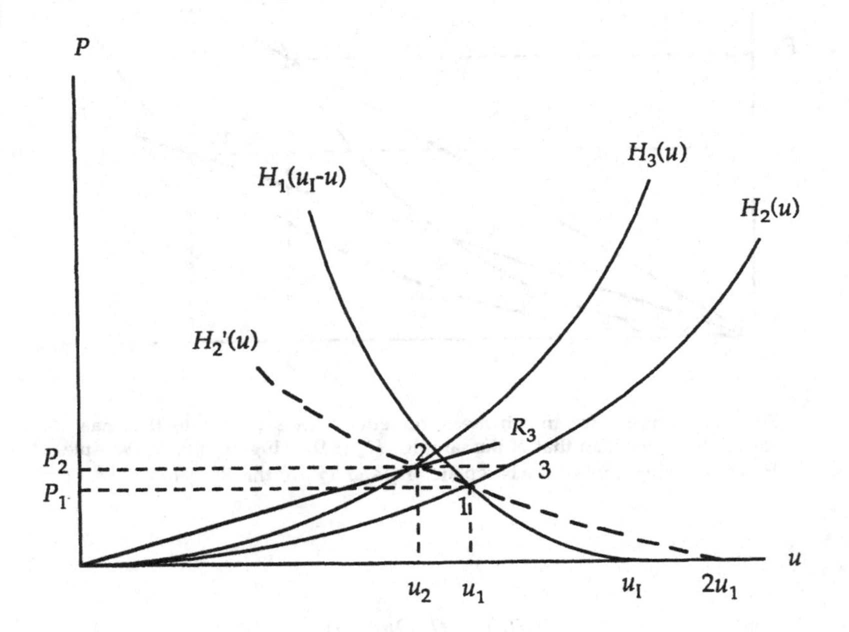

Fig. 7.5. Reshock analysis. The initial shock state is at point 1. The state produced on reshock is at point 2. Point 3 is the inferred point on the principal Hugoniot of the sample. R_3 is the Rayleigh line for the reshock.

8. Sound Speed Measurement Techniques

8.1 Edge Rarefaction Method

Consider an opaque target shocked to pressures sufficient to cause it to become self-luminous. The edges of the target are free surfaces. Figure 8.1 illustrates the geometrical arrangement.

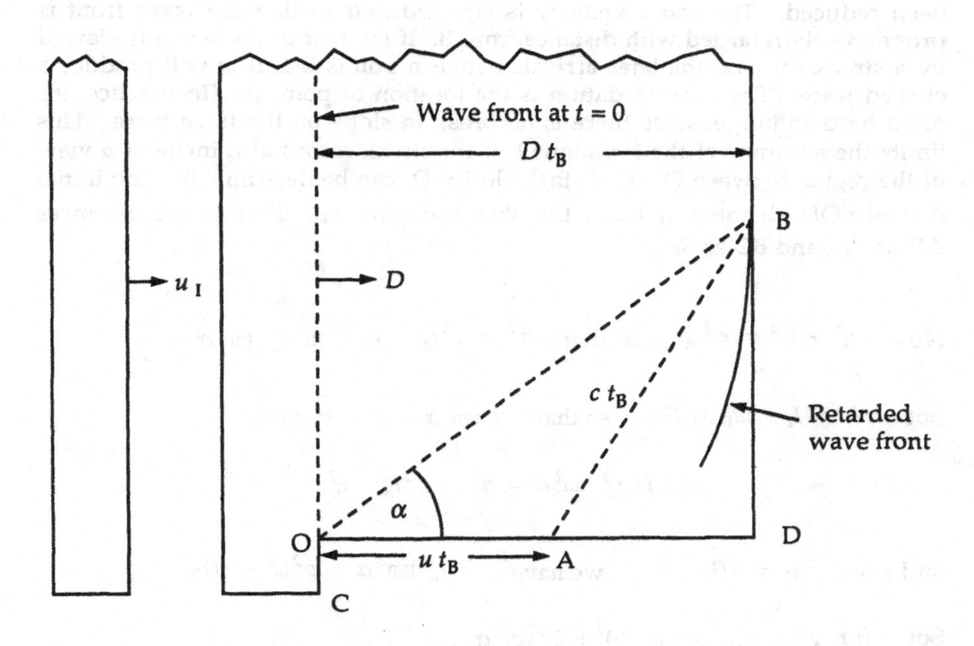

Fig. 8.1. Measurement of sound speed behind a shock using the edge rarefaction method. The target is assumed to be opaque. For a strong enough shock, when the initial shock breaks out between O and C, a flash of light occurs, signalling the time. Similarly, when the shock breaks out at the rear surface above B, the surface lights up. The light from the retarded portion of the wave occurs at later times. A streak camera can be used to determine the times and locate point B. Analysis of the geometry then yields the sound speed behind the shock.

Consider a shock that passes point O at time zero, and arrives at the rear surface of the target at time t_B (breakout time). After the shock passes O the material acquires particle velocity u, so that at time t_B, point O has advanced to point A. In the meanwhile, since the lateral surface is free, a lateral release wave with earliest arrival traveling at the sound speed c established behind the shock has propagated a distance ct_B from point A. The portions of the shock wave between points B and D have been reduced in pressure by release waves, so that the material and sound velocities have been reduced. The shock velocity is also reduced so that the wave front is progressively retarded with distance from B. If the rear of the target is viewed by a streak camera, the later arrivals between points B and D will produce a curved trace. The desired datum is the location of point B. (In practice, it's often hard to judge, since there is no break in slope on the trace there. This limits the accuracy of the technique.) If the streak record also includes a view of the region between O and C, the velocity D can be determined using initial distance OD (denoted l_1) and the observed time t_B. Denote the distances AD as l_2, and BD as h.

Now $l_2^2 + h^2 = c^2 t_B^2$, so that $h^2 = c^2 t_B^2 - l_2^2$ and $\tan\alpha = \dfrac{h}{l_1}$

but $l_1 = l_2 + u t_B = D t_B$ so that $l_1 \tan\alpha = h$ becomes

$$D^2 t_B^2 \tan^2\alpha = h^2 = c^2 t_B^2 - l_2^2$$

and since $l_2 = (D-u)t_B$ we have $D^2 t_B^2 \tan^2\alpha = c^2 t_B^2 - (D-u)^2 t_B^2$

Solve for c^2: $c^2 = (D-u)^2 + D^2 \tan^2\alpha$

Finally, we have

$$c = D\sqrt{\left(\frac{D-u}{D}\right)^2 + \tan^2\alpha} \qquad (8.1)$$

If the Hugoniot of the target material is known, the value of u can be determined from D.

8.2 Overtaking Rarefaction Method.

A second method for measuring sound speed behind the shock front has been developed in recent years. For simplicity, we will illustrate the case of symmetric impact, where a target and impactor are made of the same material. Figure 8.2 illustrates the *x-t* diagram for the situation where the target is thick enough so that the overtaking rarefaction wave from the rear of the impactor overtakes the shock in the target. The x coordinate is Lagrangian (coordinates fixed in the material) rather than in the lab frame, so that the motion of the surfaces is not shown. We will consider only the lead characteristic. Time zero is taken as the instant of impact. The shock velocity D is determined from the known Hugoniot of the materials and the impact velocity. A shock is launched in the target and a backward moving shock is launched in the impactor. At time t_1 the backward moving shock reaches the rear surface of the impactor and a release wave begins from point 1 moving at the Lagrangian velocity c^L (the velocity relative to the medium). The lead characteristic is shown.

At point 2 the lead characteristic overtakes the forward shock in the target and the shock begins to be weakened. Define R as the ratio of the depth x_2 in the target where overtake occurs to the original thickness ($-x_1$) of the impactor. We have

$$x_1 = -Dt_1, \quad x_2 = Dt_2$$

but

$$x_2 - x_1 = c^L(t_2 - t_1)$$

or

$$c^L(t_2 - t_1) = D(t_2 + t_1)$$

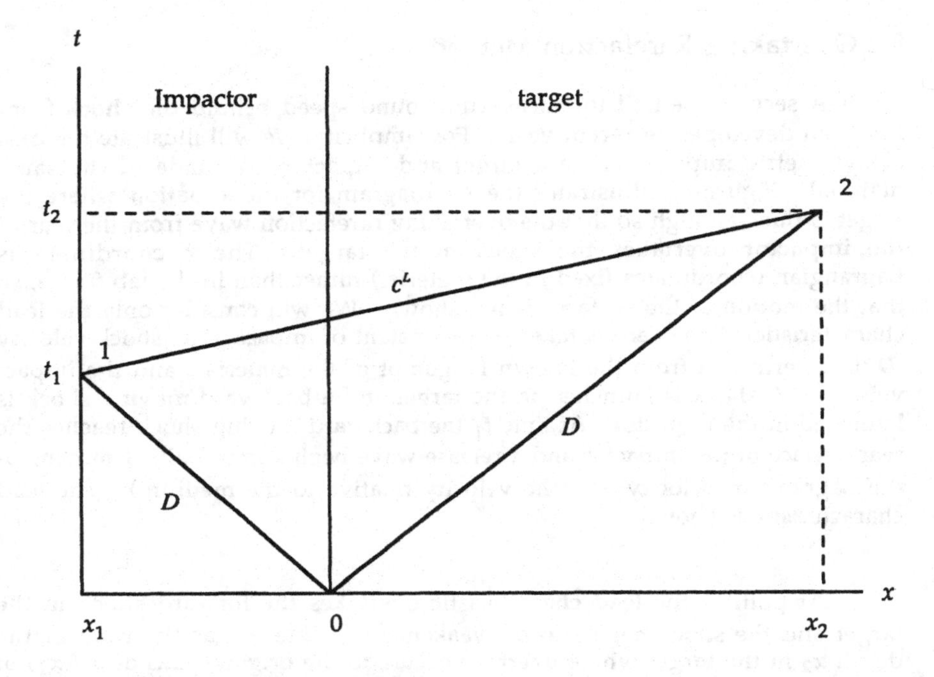

Fig. 8.2. *x-t* diagram for a symmetric collision of a thin impactor with a thick target. The *x* coordinate is Lagrangian, so that movement of the surfaces is not shown. The forward-going shocks are shown with the overtaking release wave. Overtake occurs at point 2.

thus

$$c^L = D\frac{(t_2 + t_1)}{(t_2 - t_1)}$$

Now

$$R \equiv \frac{x_2}{-x_1} = \frac{Dt_2}{Dt_1} = \frac{t_2}{t_1}$$

so that

$$c^L = D\left(\frac{R+1}{R-1}\right)$$

The density of the material behind the shock can be determined from the known Hugoniot and the measured shock strength. The sound velocity behind the shock can then be determined from the Lagrangian velocity by multiplying by the ratio of the initial density to the shocked density:

$$c = c^L\left(\frac{\rho_0}{\rho}\right)$$

The problem is thus how to determine the overtake point in order to determine R .

Consider a target thin enough for the initial shock in the target to reach the rear of the target before the overtaking release wave can reach it. The rear of the target is backed by a medium called the analyzer, that is transparent and has the property that strong shocks cause it to emit copious amounts of light from the shock front and the intensity is a strong function of the shock pressure. Figure 8.3 illustrates the corresponding *x-t* diagram. The *x* coordinate is again Lagrangian. We will assume that the Hugoniot of the backing medium is known and for the sake of the illustration, we will assume that its Hugoniot is less stiff than that of the target, so that a rarefaction wave is sent back into the target (similar analysis holds for the opposite case, but this is the choice shown in the figure.) The lead characteristic for this wave from point 2 to point 3 has the same slope but opposite sign to that of the wave from point 1 to point 3. At point 3 the waves meet and a new velocity c_2^L is established. The overtaking wave then continues at reduced velocity to point 4 on the rear face of the target. On entering the analyzer a new velocity c_3^L is established. The shock in the analyzer from point 2 moves with velocity D_A which can be calculated from D and the Hugoniot of the analyzer. At point 5 the overtaking rarefaction catches the shock and the pressure begins to drop, causing the light output from the shock in the analyzer to decrease. The time elapsed between points 2 and 5 can be recorded with a streak camera viewing the analyzer from the rear. Since the initial dimensions are known and the velocity D_A can be calculated, the point 5 is determined. Now suppose the target has steps machined on the rear surface so that a series of target thicknesses are present. For the next thickness, another set of locations for points 2, 4 and 5 exist (call them 2', 4', and 5'). The curve from point 5 to point 5' passes through point 6 and it can be shown that it is linear. The intersection of this line with the line from point 2 with slope D gives point 6, which is where overtake would have occurred if the target were infinitely thick. In practice, a series of steps are used and a least squares fit is used to determine the line through point 6. The analyzer medium is commonly a liquid such as bromoform since it conforms to the rear of the target and has good optical properties and a known Hugoniot.

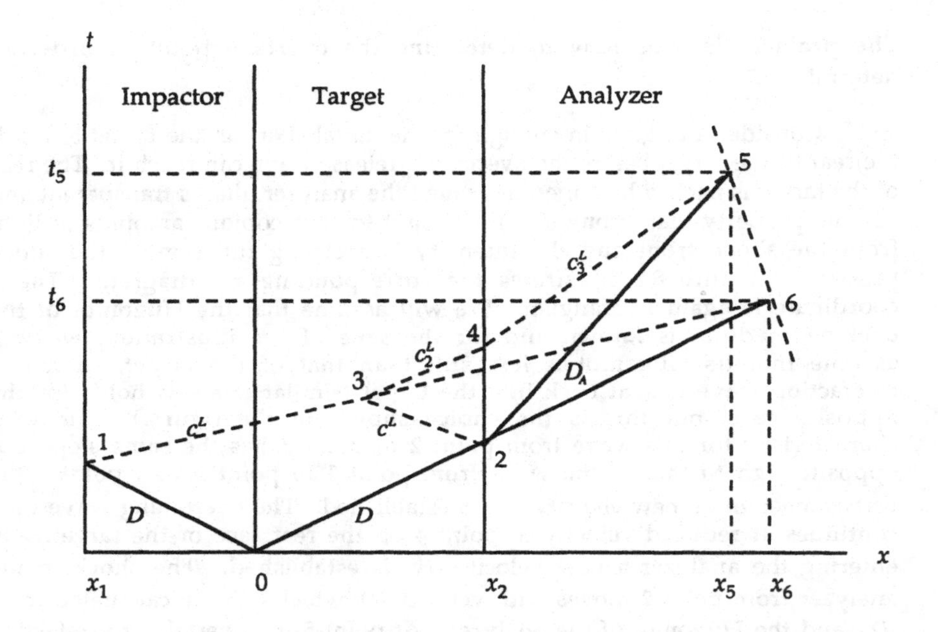

Fig. 8.3. x-t diagram for a thin impactor striking a thin target backed by an analyzer medium. The x coordinate is Lagrangian, so that movement of the surfaces is not seen. At point 5 the rarefaction wave overtakes the shock wave in the analyzer medium, causing the light intensity from the shock to decrease. Point 6 is where overtake would occur in a very thick target.

9. Release Isentrope Measurement Techniques

9.1 Overtaking Wave Method

Consider the collision of an impactor with a thick target, which we will assume is of the same material for simplicity. The target is initially at rest. Figure 9.1 shows the *x-t* diagram for the one dimensional aspects of the collision. The *x* coordinate here is in lab frame. The initial shock propagates with velocity *D*, determining its slope in the diagram. The release fan from point 1 (the rear of the impactor at the time the rear going shock arrives) does not change slope at the materials interface because the materials are the same. At point 3 the lead characteristic from point 1 overtakes the shock wave. Accordingly, the wave begins to slow down as it is attenuated. Assume that we can measure the initial velocity of the impactor and the path $x(t)$ followed by the attenuating wave, as well as the particle velocity along the path.

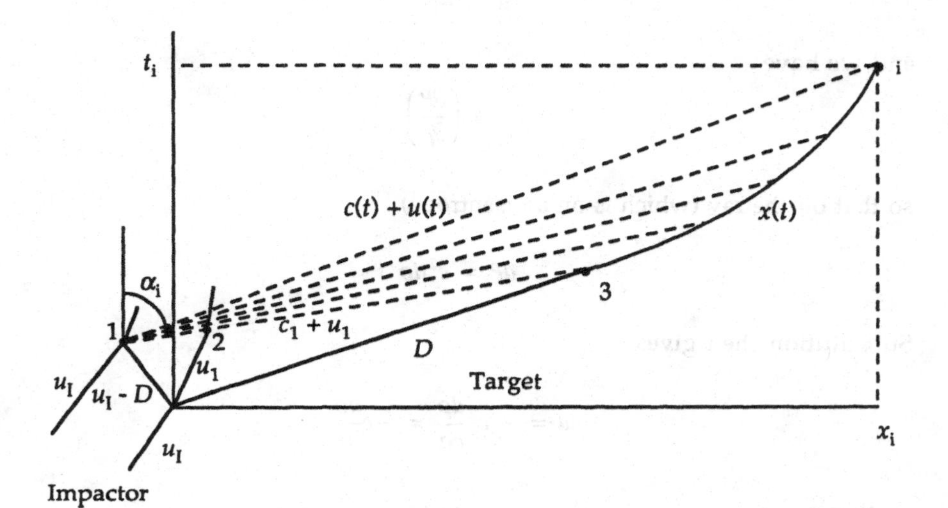

Fig. 9.1. *x-t* diagram for collision of a thin impactor with a target. The *x* coordinate is in the lab frame. The shock in the target is overtaken by the lead characteristic from the rear of the impactor at point 3. As the release fan catches the wave, it progressively slows down.

If the Hugoniot is known, the impactor velocity u_I and the Hugoniot give us D and u behind the unattenuated shock. Geometric construction using the known thickness of the impactor then gives us the triangle formed by the origin and points 1 and 3, which is formed by the forward and backward going shocks and the lead ray of the release fan. Points 1 and 3 are thus established. For an arbitrary point on the path $x(t)$ we can construct the ray from point 1, thus determining its slope. We have

$$\tan \alpha_i = c(t_i) + u(t_i)$$

But since we are separately measuring $u(t_i)$ we can subtract it to determine $c(t_i)$. We thus can determine sound speed $c(t)$ along the track $x(t)$. Since we have $c(t)$ and $u(t)$ along the track, we can eliminate time to give $c(u)$. For a given ray from point 1 to the curve $x(t)$ we have the Riemann invariant

$$du = -\frac{dP}{\rho c}$$

and we have

$$c^2 \equiv \left(\frac{\partial P}{\partial \rho}\right)_s$$

so that on the ray (which is on an isentrope)

$$dP = c^2 d\rho$$

Substitution then gives

$$du = -c^2 \frac{d\rho}{c\rho} = -c\frac{d\rho}{\rho}$$

so that

$$\frac{du}{c(u)} = -\frac{d\rho}{\rho}$$

Integration then gives $\rho(u)$. Since we now have both $c(u)$ and $\rho(u)$ we can eliminate u to give $c(\rho)$ and then

$$dP = c^2(\rho)d\rho$$

Integration of this form gives $P(\rho)$ along the track $x(t)$ or along the release isentrope. The chief limitations on this technique is that $x(t)$ and $u(t)$ are difficult to measure, and the release tail is very long. As a result, the technique would be primarily useful for the initial portions of the release. Large pressure drops call for long times. This means the Target would have to be quite thick. In order to prevent lateral release waves from destroying the one dimensional behavior, the target would also have to be very wide. This calls for large impactors which could not be accelerated to interesting velocities with available launchers.

9.2 Calibrated Release Method

Suppose we know the principal Hugoniots for a material to be studied and a variety of softer materials. Consider the collision of an impactor (whose Hugoniot need not be accurately known) with a stationary target of the material to be studied, backed with a number of softer (standard) materials. Each of the standard materials is in contact with the back of the target. Detectors are provided to determine the arrival time of the shock at the rear face of each material. Figure 9.2 shows the mechanical arrangement. Detectors 1 through 5 establish the tilt and shape of the shock wave. An arrangement of three standard materials is shown. Nine detectors are used. Detector 6 with the known step height in the baseplate is used to determine the shock velocity in the baseplate. Figure 9.3 shows the relationships in the pressure - particle velocity plane. We measure $\rho_{0x},\ldots,\rho_{03}$ and D_x, D_1, D_2, and D_3 to determine the Rayleigh lines R_x, R_1, R_2, and R_3. The equations for the Rayleigh lines are

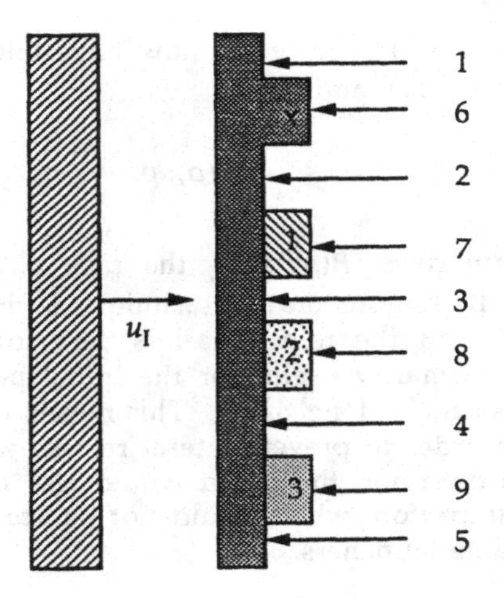

Fig. 9.2. Mechanical arrangement for calibrated release measurement of release isentropes. The target baseplate is the material to be studied. Materials 1,2, and 3 are standards with known Hugoniots. Detectors 1 through 5 establish the tilt and shape of the shock wave. Detector 6 with the known step height in the baseplate is used to determine the shock velocity in the baseplate.

$$
\begin{aligned}
P_a - P_0 &= \rho_{0x} D_x u \\
P_b - P_0 &= \rho_{01} D_1 u \\
P_c - P_0 &= \rho_{02} D_2 u \\
P_d - P_0 &= \rho_{03} D_3 u
\end{aligned}
$$

The intersection of the Rayleigh lines with the corresponding Hugoniots determine points a, b, c, and d which lie on the release isentrope of the target material extending from the shocked state at a. While in principle points b, c, and d could be determined on successive shots, allowing simpler design, the launchers available are not consistent enough to repeat the state at point a. Accuracy would thus be severely degraded. The release isentrope is determined by the equation of state of the material. Pressure and particle velocity are not particularly convenient variables. It is desirable to convert to more conventional thermodynamic variables. From the differential forms of the Riemann invariants on a release isentrope we can write

$$du = \pm \frac{dP}{\rho c}$$

where ρ and c are the density and sound speed. We can thus write

$$\left(\frac{du}{dP}\right)^2 = \frac{1}{\rho^2 c^2}$$

and since

$$c^2 = \left(\frac{\partial P}{\partial \rho}\right)_s$$

we have

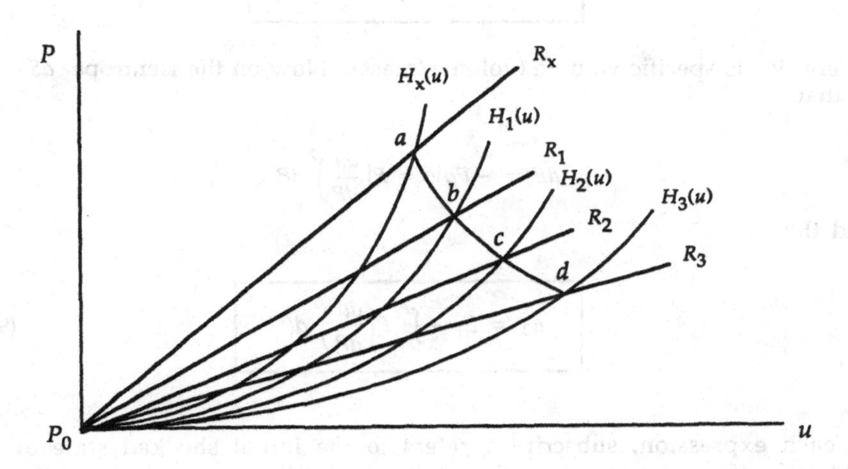

Fig. 9.3. Calibrated release measurement. The Hugoniots of the materials are shown along with the Rayleigh lines for the various shocks in the standards and sample.

$$\left(\frac{du}{dP}\right)^2 = \frac{1}{\rho^2}\left(\frac{d\rho}{dP}\right)_s$$

so that on the release isentrope we have

$$\frac{d\rho}{\rho^2} = \left(\frac{du}{dP}\right)^2 dP$$

but $V = 1/\rho$ so that

$$dV = -\left(\frac{du}{dP}\right)^2 dP$$

We can thus write

$$\boxed{V_2 = V_1 - \int_{P_1}^{P_2}\left(\frac{du}{dP}\right)^2 dP} \qquad (9.1)$$

where V is specific volume (volume/mass). Now on the isentrope $dS = 0$ so that

$$dE = -PdV = P\left(\frac{du}{dP}\right)^2 dP$$

and thus

$$\boxed{E_2 = E_1 + \int_{P_1}^{P_2} P\left(\frac{du}{dP}\right)^2 dP} \qquad (9.2)$$

In each expression, subscript 1 refers to the initial shocked state of the material. If we assume the points *a, b, c,* and *d* lie on a smooth continuous curve we can fit the points with a polynomial and evaluate the integrals analytically as running functions along the isentrope. If the form of the curve is not simple, numerical integration may be used. Let's assume a quadratic form

$$P = a_0 + a_1 u + a_2 u^2$$

for the isentrope. We have

$$\frac{dP}{du} = a_1 + 2a_2 u$$

and thus

$$\left(\frac{dP}{du}\right)^2 = \left(a_1^2 - 4a_2 a_0\right) + 4a_2 P$$

define $\xi \equiv \left(a_1^2 - 4a_2 a_0\right) + 4a_2 P$ so that $d\xi = 4a_2 dP$ or $dP = d\xi/4a_2$

$$\left(\frac{du}{dP}\right)^2 dP = \frac{1}{4a_2}\frac{d\xi}{\xi} = \frac{1}{4a_2} d \ln \xi$$

We then have

$$\int_{P_1}^{P_2} \left(\frac{du}{dP}\right)^2 dP = \frac{1}{4a_2} \ln\left(\frac{\xi_2}{\xi_1}\right)$$

where

$$\xi_1 = \left(a_1^2 - 4a_2 a_0\right) + 4a_2 P_1$$
$$\xi_2 = \left(a_1^2 - 4a_2 a_0\right) + 4a_2 P_2$$

thus

$$\boxed{V_2 = V_1 - \frac{1}{4a_2} \ln\left(\frac{\xi_2}{\xi_1}\right)} \tag{9.3}$$

Now

$$P = \frac{\xi - (a_1^2 - 4a_2 a_0)}{4a_2}$$

so that

$$P\left(\frac{du}{dP}\right)^2 dP = \frac{\xi-(a_1^2-4a_2a_0)}{4a_2}\frac{1}{\xi}\frac{d\xi}{4a_2} = \frac{1}{(4a_2)^2}\left[d\xi-\left(a_1^2-4a_2a_0\right)d\ln\xi\right]$$

and

$$\int_{P_1}^{P_2} P\left(\frac{du}{dP}\right)^2 dP = \frac{1}{(4a_2)^2}\int_{\xi_1}^{\xi_2} d\xi - \frac{(a_1^2-4a_2a_0)}{(4a_2)^2}\int_{\xi_1}^{\xi_2} d\ln\xi$$

$$= \frac{1}{(4a_2)^2}\left[\xi_2-\xi_1-\left(a_1^2-4a_2a_0\right)\ln\left(\frac{\xi_2}{\xi_1}\right)\right]$$

but

$$\frac{\xi_2-\xi_1}{(4a_2)^2} = \frac{4a_2(P_2-P_1)}{(4a_2)^2} = \frac{P_2-P_1}{4a_2}$$

so that finally, we have

$$\boxed{E_2 = E_1 + \frac{P_2-P_1}{4a_2} - \frac{(a_1^2-4a_2a_0)}{(4a_2)^2}\ln\left(\frac{\xi_2}{\xi_1}\right)} \qquad (9.4)$$

where

$$\xi_1 = \left(a_1^2-4a_2a_0\right)+4a_2P_1$$
$$\xi_2 = \left(a_1^2-4a_2a_0\right)+4a_2P_2$$

The step of first fitting a polynomial to the data, followed by differentiation of the polynomial to construct the integrals is used, since numerical differentiation is notoriously noisy and would be absurd for so few points. If the resulting integrand is too complex for analytic calculation, numerical methods may be readily used, since integration is a smoothing process.

If a kink is present in the isentrope, such as at the liquid-vapor coexistence boundary and appears in the data, separate polynomials can be fitted on either side of the kink. The integrals then break into two parts.

10. Measurement of the Grüneisen Coefficient

The Grüneisen coefficient may be determined experimentally by making shock wave measurements on samples of different initial densities. Assume we have Hugoniots corresponding to two different initial densities. One can estimate γ_G at a given density common to both Hugoniots. Since we have

$$\gamma_G \equiv V\left(\frac{\partial P}{\partial E}\right)_V$$

we can make the approximation

$$\gamma_G \cong V\left(\frac{\Delta P}{\Delta E}\right)_V$$

From the Hugoniot energy equation we have

$$E - E_0 = \frac{1}{2}(P_H + P_0)(V_0 - V)$$

Assume that we can neglect E_0 and P_0 , so that

$$E \cong \frac{1}{2}P_H(V_0 - V)$$

Choose a particular specific volume (corresponding to the fact that the derivative of pressure with respect to energy is at constant volume) and determine the corresponding pressure on each Hugoniot, then we have

$$(E_2 - E_1) = \frac{1}{2}P_{H_2}(V_{02} - V) - \frac{1}{2}P_{H_1}(V_{01} - V) = \frac{1}{2}\left(P_{H_2}V_{02} - P_{H_1}V_{01}\right) - \frac{1}{2}V\left(P_{H_2} - P_{H_1}\right)$$

Now

$$\gamma_G \cong V\left(\frac{P_{H_2} - P_{H_1}}{E_2 - E_1}\right)_V$$

so that

$$E_2 - E_1 = \frac{V\left(P_{H_2} - P_{H_1}\right)}{\gamma_G}$$

thus

$$\frac{V\left(P_{H_2} - P_{H_1}\right)}{\gamma_G} = \frac{1}{2}\left(P_{H_2} V_{02} - P_{H_1} V_{01}\right) - \frac{1}{2} V\left(P_{H_2} - P_{H_1}\right)$$

or

$$\frac{1}{\gamma_G} = \frac{1}{2V}\left(\frac{P_{H_2} V_{02} - P_{H_1} V_{01}}{P_{H_2} - P_{H_1}}\right) - \frac{1}{2}$$

This gives the apparent γ_G for the particular specific volume selected. The analysis can be repeated for the range of densities shared by both Hugoniots to see the apparent volume dependence of γ_G. In practice measurement of the Grüneisen coefficient by these methods suffers from nonuniformity in sample density and the resulting values are quite noisy.

Consider a Hugoniot described by a linear D, u fit. We have

$$D = c_0 + su$$
$$P = \rho_0 D u$$
$$\rho = \rho_0 \left(\frac{D}{D-u}\right) = \rho_0 \frac{c_0 + su}{\left[c_0 + (s-1)u\right]}$$

Solve for u and then D to obtain

$$u = \frac{c_0 (\rho - \rho_0)}{\left[\rho_0 s - \rho(s-1)\right]}$$
$$D = \frac{c_0 \rho}{\left[\rho_0 s - \rho(s-1)\right]}$$

We can then substitute in the expression for the pressure to obtain pressure as a function of the density.

$$P = \frac{\rho_0 c_0^2 \rho(\rho - \rho_0)}{[\rho_0 s - \rho(s-1)]^2}$$

We can then take the fits to the Hugoniots for the two different initial densities, and evaluate the pressure on each for the given density. The results can then be substituted in

$$\frac{1}{\gamma_G} = \frac{\rho}{2}\left[\frac{P_2/\rho_{02} - P_1/\rho_{01}}{P_2 - P_1}\right] - \frac{1}{2}$$

to obtain the corresponding Grüneisen coefficient γ_G.

As an example, consider 2024 Al alloy, for which the following data apply[9].

$$\rho_0 = 2.785 \text{ g/cm}^3,\ c_0 = 0.5328 \text{ cm/}\mu\text{s},\ s = 1.338$$
$$\rho_{01} = 2.543 \text{ g/cm}^3,\ c_0 = 0.5091 \text{ cm/}\mu\text{s},\ s_1 = 1.369$$
$$\rho_{02} = 2.300 \text{ g/cm}^3,\ c_{02} = 0.4842 \text{ cm/}\mu\text{s},\ s_2 = 1.400$$

Consider a density of $\rho = 3.7 \text{ g/cm}^3$. The pressures on the Hugoniots are 0.43669 MBar, 0.63013 MBar, and 0.92259 MBar, respectively. If we calculate γ_G using the first and third Hugoniots we obtain $\gamma_G = 2.324$. If we use the second and third Hugoniots we obtain $\gamma_G = 2.128$. If we use the first and second Hugoniots we obtain $\gamma_G = 2.701$.

11. Calculation of the Principal Hugoniot For an Analytic Equation of State

11.1 Method of Calculation

Assume $u_0 = 0$, $P_0 = 0$ (material at rest at ambient pressure)

$$u = \sqrt{(P-P_0)(V_0-V)}, \text{ so that } \frac{u^2}{2} = \frac{1}{2}(P-P_0)(V_0-V)$$

but

$$E-E_0 = \frac{1}{2}(P+P_0)(V_0-V)$$

$$\frac{u^2}{2}+(E-E_0) = \frac{V_0-V}{2}\left[P-P_0+P+P_0\right] = P(V_0-V)$$

but for $\quad P_0 = 0, \ \frac{u^2}{2} = \frac{P(V_0-V)}{2} \quad$ or $\quad P(V_0-V) = u^2$.

then $\quad \frac{u^2}{2}+(E-E_0) = u^2 \quad$ so that $\quad E-E_0 = \frac{u^2}{2}$.

Now $\varepsilon = \rho_0 E \quad$ hence $\quad \varepsilon-\varepsilon_0 = \rho_0\frac{u^2}{2}$

Most codes choose $\varepsilon_0 = 0$, so we will drop it for our example. Thus on the Hugoniot, we have $\varepsilon \equiv \rho_0\frac{u^2}{2}$. Substitute this in the equation of state:

$$P_H(\mu,u) = P_{EOS}(\mu,u)\Big|_{\varepsilon = \rho_0\frac{u^2}{2}} \qquad \text{where } \mu = \frac{\rho}{\rho_0}-1.$$

Now conservation of mass gives $\rho(D-u) = \rho_0(D-u_0)$, or since $u_0 = 0$,

$$\frac{\rho}{\rho_0} = \frac{D}{D-u} = \mu+1\,,\ \mu = \frac{u}{D-u}\,,\ D = \left(\frac{\mu+1}{\mu}\right)u$$

Now $P = \rho_0 D u$ (conservation of momentum) so that

$$P = \rho_0\left(\frac{\mu+1}{\mu}\right)u^2$$

We thus have

$$P_{EOS}(\mu, u)\Big|_{\varepsilon = \rho_0\frac{u^2}{2}} - \rho_0\left(\frac{\mu+1}{\mu}\right)u^2 = 0$$

Choose a value of μ and solve this for u, then

$$D = \left(\frac{\mu+1}{\mu}\right)u$$

This gives the point (D, u) corresponding to a given μ. Calculate a set of points (D, u) for a range of μ values, then fit the collection of points to describe the Hugoniot. In general, for complicated EOS forms, root finding methods will be required to determine u corresponding to a given μ. For forms linear in the internal energy, an analytic solution can be found. The following cases are examples. The specific EOS models will be more fully developed in the next chapter.

11.2 PUFF Equation of State

$$P = a_1\mu + a_2\mu^2 + a_3\mu^3 + b_1(\mu+1)\varepsilon$$

$$P_{EOS}(\mu, \varepsilon)\Big|_{\varepsilon = \rho_0\frac{u^2}{2}} = a_1\mu + a_2\mu^2 + a_3\mu^3 + b_1(\mu+1)\rho_0\frac{u^2}{2}$$

We have

$$a_1\mu + a_2\mu^2 + a_3\mu^3 + \rho_0(\mu+1)\left[\frac{b_1}{2} - \frac{1}{\mu}\right]u^2 = 0$$

or

$$u = \sqrt{\frac{a_1\mu + a_2\mu^2 + a_3\mu^3}{\rho_0(\mu+1)(1/\mu + b_1/2)}}, \quad D = \left(\frac{\mu+1}{\mu}\right)u, \quad P = \rho_0 D u$$

11.3 Polynomial Equation of State

$$P = c_0 + c_1\mu + c_2\mu^2 + c_3\mu^3 + \left(c_4 + c_5\mu + c_6\mu^2\right)\varepsilon$$

$$u = \sqrt{\frac{c_0 + c_1\mu + c_2\mu^2 + c_3\mu^3}{\rho_0\left[(\mu+1)/\mu - (c_4 + c_5\mu + c_6\mu^2)/2\right]}}$$

$$D = \left(\frac{\mu+1}{\mu}\right)u, \quad P = \rho_0 D u$$

11.4 Grüneisen Equation of State

$$P = \frac{\rho_0 c_0^2 \mu\left[1 + (1-\gamma_0/2)\mu - a\mu^2/2\right]}{\left[1 - (S_1 - 1)\mu - S_2\frac{\mu^2}{(\mu+1)} - S_3\frac{\mu^3}{(\mu+1)^2}\right]^2} + (\gamma_0 + a\mu)\varepsilon$$

Define

$$\omega(\mu) \equiv \frac{c_0^2 \mu\left[1 + (1-\gamma_0/2)\mu - a\mu^2/2\right]}{\left[1 - (S_1 - 1)\mu - S_2\frac{\mu^2}{(\mu+1)} - S_3\frac{\mu^3}{(\mu+1)^2}\right]^2}$$

then

$$P = \rho_0\,\omega(\mu) + (\gamma_0 + a\mu)\varepsilon$$

On the Hugoniot we have

$$\rho_0\,\omega(\mu) + (\gamma_0 + a\mu)\rho_0\frac{u^2}{2} - \rho_0\left(\frac{\mu+1}{\mu}\right)u^2 = 0$$

thus

$$\omega(\mu) = \left[\left(\frac{\mu+1}{\mu}\right) - \frac{(\gamma_0 + a\mu)}{2}\right]u^2$$

or

$$u = \sqrt{\frac{\omega(\mu)}{\left[\left(\frac{\mu+1}{\mu}\right) - \frac{(\gamma_0+a\mu)}{2}\right]}}$$

$$D = \left(\frac{\mu+1}{\mu}\right)u\,, \quad P = \rho_0 D u$$

11.5 Appy Equation of State

$$P = \rho_0\, \omega(\mu) + \gamma_0\,(\mu)\varepsilon$$

where

$$\omega(\mu) = \frac{a^2\mu(\mu+1)(1-\gamma_0\,\mu/2)}{\left[(b-1)\mu-1\right]^2}\,, \quad \gamma_0\,(\mu) = \frac{(g_0 - g_1\,\mu)}{1+g_2\,\mu+g_3\,\mu^2}$$

$$g_0 = 2b-1\,, \quad g_1 = \frac{3}{4}g_0^2 - \frac{1}{3}g_0 - 0.6388889$$

$$g_2 = g_0\,, \quad g_3 = g_0^2 - b^2$$

On the Hugoniot we have

$$\rho_0\, \omega(\mu) + \gamma_0\, \rho_0 \frac{u^2}{2} - \rho_0 \left(\frac{\mu+1}{\mu}\right)u^2 = 0$$

so that

$$u = \sqrt{\frac{\omega(\mu)}{\left[\left(\frac{\mu+1}{\mu}\right) - \frac{\gamma_0}{2}\right]}}$$

$$D = \left(\frac{\mu+1}{\mu}\right)u\,, \quad P = \rho_0 D u$$

11.6 Ratio of Polynomials Equation of State

$$P(\mu,\varepsilon) = (1+\mu)\frac{\omega(\mu,\varepsilon)}{\psi(\mu,\varepsilon)}$$

where

$$\omega(\mu,\varepsilon) \equiv F_1(\mu)+F_2(\mu)\varepsilon+F_3(\mu)\varepsilon^2+F_4(\mu)\varepsilon^3$$
$$\psi(\mu,\varepsilon) \equiv F_5(\mu)+F_6(\mu)\varepsilon+F_7(\mu)\varepsilon^2$$

and

$$F_1(\mu) = a_1+a_2\mu+a_3\mu^2+a_4\mu^3$$
$$F_2(\mu) = a_5+a_6\mu+a_7\mu^2+a_8\mu^3$$
$$F_3(\mu) = a_9+a_{10}\mu+a_{11}\mu^2+a_{12}\mu^3$$
$$F_4(\mu) = a_{13}+a_{14}\mu+a_{15}\mu^2+a_{16}\mu^3$$
$$F_5(\mu) = a_{17}+a_{18}\mu+a_{19}\mu^2+a_{20}\mu^3$$
$$F_6(\mu) = a_{21}+a_{22}\mu+a_{23}\mu^2+a_{24}\mu^3$$
$$F_7(\mu) = a_{25}+a_{26}\mu+a_{27}\mu^2+a_{28}\mu^3$$

Now define

$$f(\mu,\varepsilon) \equiv P(\mu,\varepsilon)\Big|_{\varepsilon=\rho_0\frac{u^2}{2}} - \rho_0\left(\frac{\mu+1}{\mu}\right)u^2$$

$$f'(\mu,u) = \left(\frac{\partial}{\partial u}\right)_\mu P(\mu,\varepsilon)\Big|_{\varepsilon=\rho_0\frac{u^2}{2}} - 2\rho_0\left(\frac{\mu+1}{\mu}\right)u$$

but

$$\left(\frac{\partial}{\partial u}\right)_\mu P(\mu,\varepsilon)\Big|_{\varepsilon=\rho_0\frac{u^2}{2}} = \left(\frac{\partial P}{\partial \varepsilon}\right)_\mu \frac{d\varepsilon}{du}\Big|_{\varepsilon=\rho_0\frac{u^2}{2}}$$

where $\dfrac{d\varepsilon}{du} = \rho_0 u$ and

$$\left(\frac{\partial P}{\partial \varepsilon}\right)_\mu \Bigg|_{\varepsilon=\rho_0\frac{u^2}{2}} = (1+\mu)\left[\frac{1}{\psi}\left(\frac{\partial \omega}{\partial \varepsilon}\right)_\mu - \frac{1}{\psi^2}\left(\frac{\partial \psi}{\partial \varepsilon}\right)_\mu\right]\Bigg|_{\varepsilon=\rho_0\frac{u^2}{2}}$$

so that

$$\left(\frac{\partial}{\partial u}\right)_\mu P(\mu,\varepsilon)\Bigg|_{\varepsilon=\rho_0\frac{u^2}{2}} = \rho_0 u(1+\mu)\left[\frac{1}{\psi}\left(\frac{\partial \omega}{\partial \varepsilon}\right)_\mu - \frac{1}{\psi^2}\left(\frac{\partial \psi}{\partial \varepsilon}\right)_\mu\right]\Bigg|_{\varepsilon=\rho_0\frac{u^2}{2}}$$

Now

$$\left(\frac{\partial \omega}{\partial \varepsilon}\right)_\mu \equiv \omega' = F_2(\mu)+2F_3(\mu)\varepsilon+3F_4(\mu)\varepsilon^2$$

$$\left(\frac{\partial \psi}{\partial \varepsilon}\right)_\mu \equiv \psi' = F_6(\mu)+2F_7(\mu)\varepsilon$$

Consider using Newton's method:

$$u_2 = u_1 - \frac{f(u_1)}{f'(u_1)}$$

$$f(u) = (1+\mu)\frac{\omega}{\psi}\Bigg|_{\varepsilon=\rho_0\frac{u^2}{2}} - \rho_0\left(\frac{\mu+1}{\mu}\right)u^2$$

$$f'(u) = \rho_0 u(1+\mu)\left[\frac{\omega'}{\psi} - \frac{\psi'}{\psi^2}\right]\Bigg|_{\varepsilon=\rho_0\frac{u^2}{2}} - 2\rho_0\left(\frac{\mu+1}{\mu}\right)u$$

We then can obtain

$$\frac{f(u)}{f'(u)} = \frac{(\omega\mu/\rho_0\psi)-u^2}{u\mu\left[\left(\frac{\omega'}{\psi}-\frac{\psi'}{\psi^2}\right)\right]-2u}$$

and

$$u-\frac{f(u)}{f'(u)} = u-\frac{(\omega\mu/\rho_0\psi)-u^2}{u\left[\mu\left(\frac{\omega'}{\psi}-\frac{\psi'}{\psi^2}\right)-2\right]}\Bigg|_{\varepsilon=\rho_0\frac{u^2}{2}}$$

Procedure:

1. For a given μ, calculate $F_1(\mu),\ldots, F_7(\mu)$

2. Choose an initial trial root u_1 , then $\varepsilon = \rho_0 \dfrac{u_1^2}{2}$

3. Calculate $\omega, \psi, \omega', \psi'$.

4. Calculate $u_2 = u_1 - \dfrac{f(u_1)}{f'(u_1)}$

5. Repeat steps 2 - 4 with the successive guesses for u until convergence occurs, then

6. $D = \left(\dfrac{\mu+1}{\mu}\right)u$, $P = \rho_0 D u$

If Newton's method does not converge, then either cut and try methods, or slope-independent methods such as interval halving may be used.

12. Equation of State Modeling

12.1 Mie-Grüneisen Equation of State

12.1.1 Derivation

Assumptions:

(1) Assume the thermal energy of a crystal is described adequately as the sum of the energies of a collection of simple harmonic oscillators, whose frequencies ν_i are functions of volume only.

(2) Neglect electronic contributions to the total internal energy E of the crystal. Let $\phi(V)$ be the potential energy of the crystal with N atoms and total volume V.

Thermodynamic Preliminaries:

A system with specified mean internal energy E is described by the canonical ensemble, where the probability density is given by

$$P_n = \frac{e^{-\beta E_n}}{\sum_n e^{-\beta E_n}}, \quad \beta \equiv \frac{1}{kT}$$

and the sum is over all states. The partition function is

$$Z \equiv \sum_n e^{-\beta E_n} \tag{12.1.1}$$

The entropy of the system is given by

$$S = k(\ln Z + \beta E) \tag{12.1.2}$$

We have

$$E = \frac{\sum_n E_n e^{-\beta E_n}}{\sum_n e^{-\beta E_n}} = -\frac{\partial}{\partial \beta} \ln Z \tag{12.1.3}$$

The Helmholtz free energy is given by

$$F = E - TS = E - \frac{1}{\beta}(\ln Z + \beta E)$$

so that

$$F = -\frac{1}{\beta} \ln Z \tag{12.1.4}$$

Now consider the harmonic oscillator, for which $E_n = (n+1/2)h\nu$

$$Z = \sum_{n=0}^{\infty} e^{-\beta(n+1/2)h\nu} = e^{-\frac{\beta h\nu}{2}} \sum_{n=0}^{\infty} e^{-n\beta h\nu}$$

But $\quad \displaystyle\sum_{n=0}^{\infty} x^n = \frac{1}{1-x}$, so that $\displaystyle\sum_{n=0}^{\infty} e^{-n\beta h\nu} = \frac{1}{1-e^{-\beta h\nu}}$

and thus

$$Z = \frac{e^{-\frac{\beta h\nu}{2}}}{1-e^{-\beta h\nu}} = \frac{1}{e^{+\frac{\beta h\nu}{2}} - e^{-\frac{\beta h\nu}{2}}}$$

or

$$Z = \frac{1}{2\sinh\left(\frac{\beta h\nu}{2}\right)} \tag{12.1.5}$$

Now therefore, we have

$$\ln Z = -\ln\left[2\sinh\left(\frac{\beta h\nu}{2}\right)\right]$$

Then

$$\frac{\partial \ln Z}{\partial \beta} = \frac{-1}{2\sinh\left(\frac{\beta h\nu}{2}\right)} \cdot 2\left(\frac{h\nu}{2}\right)\cosh\left(\frac{\beta h\nu}{2}\right) = -\frac{h\nu}{2}\coth\left(\frac{\beta h\nu}{2}\right)$$

so that

$$\boxed{E = \frac{h\nu}{2}\coth\left(\frac{\beta h\nu}{2}\right)} \tag{12.1.6}$$

Now

$$\coth x = \frac{e^x + e^{-x}}{e^x - e^{-x}} = 1 + \frac{2e^{-x}}{e^x - e^{-x}} = 1 + \frac{2}{e^{2x} - 1}$$

so that

$$\boxed{E = \frac{h\nu}{2} + \frac{h\nu}{e^{\beta h\nu} - 1}} \tag{12.1.7}$$

Now $\qquad Z = \dfrac{1}{e^{\frac{\beta h\nu}{2}} - e^{-\frac{\beta h\nu}{2}}}$ so that

$$\ln Z = -\ln\left[e^{\frac{\beta h\nu}{2}} - e^{-\frac{\beta h\nu}{2}}\right] = -\ln\left[e^{\frac{\beta h\nu}{2}}\left(1 - e^{-\beta h\nu}\right)\right] = -\frac{\beta h\nu}{2} - \ln\left(1 - e^{-\beta h\nu}\right)$$

and the Helmholtz free energy is given by

$$\boxed{F = \frac{h\nu}{2} + \frac{1}{\beta}\ln\left(1 - e^{-\beta h\nu}\right)} \tag{12.1.8}$$

We now consider a crystal of N atoms and total volume V modeled by harmonic oscillators with frequencies ν_i and add the potential energy term $\phi(V)$ to the energy of the oscillators.

$$E = \phi(V) + \sum_{i=1}^{3N} \left\{ \frac{h\nu_i}{2} + \frac{h\nu_i}{e^{h\nu_i/kT} - 1} \right\} \tag{12.1.9}$$

where the sum is over the 3N normal modes of the crystal. The Helmholtz free energy for the system is given by

$$F = \phi(V) + \sum_{i=1}^{3N} \frac{h\nu_i}{2} + kT \sum_{i=1}^{3N} \ln\left(1 - e^{-h\nu_i/kT}\right) \tag{12.1.10}$$

The total pressure is then given by $P = -\left(\frac{\partial F}{\partial V}\right)_T$ and since

$$\phi = \phi(V), \left(\frac{\partial \phi}{\partial V}\right)_T = \frac{d\varphi}{dT}$$

we have

$$-\left(\frac{\partial F}{\partial V}\right)_T = -\frac{d\phi}{dT} - \sum_{i=1}^{3N} \frac{h}{2}\left(\frac{\partial \nu_i}{\partial V}\right)_T - kT \sum_{i=1}^{3N} \left(\frac{\partial}{\partial V}\right)_T \ln\left[1 - e^{-h\nu_i/kT}\right]$$

but

$$\left(\frac{\partial}{\partial V}\right)_T \ln\left[1 - e^{-h\nu_i/kT}\right] = \frac{-e^{-h\nu_i/kT}}{\left(1 - e^{-h\nu_i/kT}\right)} \left(\frac{\partial}{\partial V}\right)_T \left(-\frac{h\nu_i}{kT}\right) = \frac{h}{kT} \frac{1}{\left(e^{h\nu_i/kT} - 1\right)} \left(\frac{\partial \nu_i}{\partial V}\right)_T$$

and thus

$$-\left(\frac{\partial F}{\partial V}\right)_T = -\frac{d\phi}{dT} - \sum_{i=1}^{3N} \frac{h}{2}\left(\frac{\partial \nu_i}{\partial V}\right)_T - h \sum_{i=1}^{3N} \frac{1}{\left(e^{h\nu_i/kT} - 1\right)} \left(\frac{\partial \nu_i}{\partial V}\right)_T \tag{12.1.11}$$

Now define the coefficients

$$\boxed{\gamma_i \equiv -\frac{d\ln \nu_i}{d\ln V} = -\frac{V}{\nu_i}\left(\frac{d\nu_i}{dV}\right)} \tag{12.1.12}$$

Since we have assumed the ν_i are functions of volume only, we can write

$$\left(\frac{\partial \nu_i}{\partial V}\right)_T = \frac{d\nu_i}{dV} = -\frac{\nu_i}{V}\gamma_i$$

and substitute in relation (12.1.11) to give

$$-\left(\frac{\partial F}{\partial V}\right)_T = -\frac{d\phi}{dT} + \sum_{i=1}^{3N}\frac{h\,\nu_i}{2\,V}\gamma_i + h\sum_{i=1}^{3N}\frac{1}{\left(e^{h\,\nu_i/kT}-1\right)}\frac{\nu_i}{V}\gamma_i$$

or

$$P = -\frac{d\phi}{dV} + \frac{1}{V}\sum_{i-1}^{3N}\gamma_i\left\{\frac{h\nu_i}{2} + \frac{h\nu_i}{\left(e^{h\,\nu_i/kT}-1\right)}\right\} \qquad (12.1.13)$$

We now make one further assumption, that all the γ_i are equal, so that they can be factored out: $\gamma_i \equiv \gamma_G$.

$$P = -\frac{d\phi}{dV} + \frac{\gamma_G}{V}\sum_{i=1}^{3N}\left\{\frac{h\nu_i}{2} + \frac{h\nu_i}{\left(e^{h\nu_i/kT}-1\right)}\right\} \qquad (12.1.14)$$

Now the vibrational part of the internal energy is $E_{vib} = E - \phi(V)$. From relation (12.1.9) we have

$$E_{vib} = \sum_{i-1}^{3N}\left\{\frac{1}{2}h\nu_i + \frac{h\nu_i}{\left(e^{h\nu_i/kT}-1\right)}\right\}$$

Substituting in relation (12.1.14) we have

$$P = -\frac{d\phi}{dV} + \frac{\gamma_G}{V}E_{vib} \qquad (12.1.15)$$

Now consider the zero degree isotherm where

$$P_k \equiv P(V)\Big|_{0°K} = -\frac{d\phi}{dV} + \frac{\gamma_G}{V}\lim_{T\to 0} E_{vib} \qquad (12.1.16)$$

As $T \to 0$, $\frac{h\nu_i}{kT} \to \infty$, $e^{h\nu_i/kT} \to \infty$, $\frac{h\nu_i}{\left(e^{h\nu_i/kT}-1\right)} \to 0$ so that

$$\lim_{T\to 0} E_{vib} = \sum_{i=1}^{3N} \frac{1}{2} h\nu_i \qquad (12.1.17)$$

Substituting in relation (12.1.16), we obtain

$$\boxed{P_k = -\frac{d\phi}{dV} + \frac{\gamma_G}{V}\sum_{i=1}^{3N}\frac{1}{2}h\nu_i} \qquad (12.1.18)$$

Similarly, $E_k \equiv E(V)\Big|_{0°K} = \phi(V) + \lim_{T\to 0} E_{vib}$, so that

$$\boxed{E_k = \phi(V) + \sum_{i=1}^{3N}\frac{1}{2}h\nu_i} \qquad (12.1.19)$$

Now consider

$$P_k - \frac{\gamma_G}{V}E_k = -\frac{d\phi}{dV} + \frac{\gamma_G}{V}\sum_{i=1}^{3N}\frac{1}{2}h\nu_i - \frac{\gamma_G}{V}\phi(V) - \frac{\gamma_G}{V}\sum_{i=1}^{3N}\frac{1}{2}h\nu_i$$

or

$$P_k - \frac{\gamma_G}{V}E_k = -\frac{d\phi}{dV} - \frac{\gamma_G}{V}\phi(V) \qquad (12.1.20)$$

Now in relation (12.1.15) we can substitute $E_{vib} = E - \phi(V)$ so that

$$P = -\frac{d\phi}{dV} + \frac{\gamma_G}{V}(E - \phi(V))$$

or

$$P = -\frac{d\phi}{dV} + \frac{\gamma_G}{V}E - \frac{\gamma_G}{V}\phi(V) \tag{12.1.21}$$

Subtracting relation (12.1.20) from relation (12.1.21) gives

$$\boxed{P - P_k = \frac{\gamma_G}{V}(E - E_k)} \tag{12.1.22}$$

Now on the Hugoniot, we may write

$$P_H - P_k = \frac{\gamma_G}{V}(E_H - E_k) \tag{12.1.23}$$

Subtracting relation (12.1.23) from relation (12.1.22) we have

$$\boxed{P - P_H = \frac{\gamma_G}{V}(E - E_H)} \tag{12.1.24}$$

This is useful for calculating states off the Hugoniot relative to measured states on the Hugoniot. Now differentiate relation (12.1.22) with respect to energy at constant volume:

$$\left(\frac{\partial P}{\partial E}\right)_V - \left(\frac{\partial P_k}{\partial E}\right)_V = \frac{1}{V}\left(\frac{\partial \gamma_G}{\partial E}\right)_V [E - E_k] + \frac{\gamma_G}{V}\left[1 - \left(\frac{\partial E_k}{\partial E}\right)_V\right]$$

But since $\gamma_G \equiv \gamma_G(V)$, $\left(\frac{\partial \gamma_G}{\partial E}\right)_V = 0$, and since $P_k = P(V)|_{0°K}$, $E_k = E(V)|_{0°K}$ we have $\left(\frac{\partial P_k}{\partial E}\right)_V = 0$, $\left(\frac{\partial E_k}{\partial E}\right)_V = 0$ and thus the result reduces to

$$\left(\frac{\partial P}{\partial E}\right)_V = \frac{\gamma_G}{V}$$

or

$$\boxed{\gamma_G = V\left(\frac{\partial P}{\partial E}\right)_V} \qquad (12.1.25)$$

Now let's derive some of the numerous equivalent forms. From thermodynamics,

$$c_v \equiv \left(\frac{\partial E}{\partial T}\right)_V \text{, therefore } (\partial E)_v = c_v\,(\partial T)_v \,.$$

Substitution in relation (12.1.25) gives

$$\gamma_G = V\frac{(\partial P)_v}{(\partial E)_v} = \frac{V}{c_v}\left(\frac{\partial P}{\partial T}\right)_V$$

so that

$$\boxed{\gamma_G = \frac{V}{c_v}\left(\frac{\partial P}{\partial T}\right)_V} \qquad (12.1.26)$$

Now $\left(\frac{\partial P}{\partial T}\right)_V\left(\frac{\partial V}{\partial P}\right)_T\left(\frac{\partial T}{\partial V}\right)_P \equiv -1$ so that $\left(\frac{\partial P}{\partial T}\right)_V = -\left(\frac{\partial P}{\partial V}\right)_T\left(\frac{\partial V}{\partial T}\right)_P$ and thus

$$\boxed{\gamma_G = -\frac{V}{c_V}\left(\frac{\partial P}{\partial V}\right)_T\left(\frac{\partial V}{\partial T}\right)_P} \qquad (12.1.27)$$

Now from thermodynamics, $\left(\frac{\partial P}{\partial V}\right)_S \equiv -\frac{c_p}{\kappa_T c_V V}$, where $\kappa_T = -\frac{1}{V}\left(\frac{\partial V}{\partial P}\right)_T$,

therefore $-\kappa_T V = \left(\frac{\partial V}{\partial P}\right)_T$, and $-\frac{1}{\kappa_T V} = \left(\frac{\partial P}{\partial V}\right)_T$, then $\left(\frac{\partial P}{\partial V}\right)_S = \frac{c_P}{c_v}\left(\frac{\partial P}{\partial V}\right)_T$,

and thus

$$\frac{1}{c_P}\left(\frac{\partial P}{\partial V}\right)_S \equiv \frac{1}{c_v}\left(\frac{\partial P}{\partial V}\right)_T \tag{12.1.28}$$

Substituting in relation (12.1.27), we have

$$\gamma_G = -\frac{V}{c_P}\left(\frac{\partial P}{\partial V}\right)_S\left(\frac{\partial V}{\partial T}\right)_P \tag{12.1.29}$$

The TdS equations from thermodynamics give

$$TdS \equiv c_v\,dT + \frac{\beta T}{\kappa_T}dV \tag{12.1.30}$$

where $\beta = \frac{1}{V}\left(\frac{\partial V}{\partial T}\right)_T, \quad \kappa_T = -\frac{1}{V}\left(\frac{\partial V}{\partial P}\right)_T$

Divide relation (12.1.30) by $c_v T$ to obtain

$$\frac{dS}{c_v} = \frac{dT}{T} + \frac{\beta}{c_v\,\kappa_T}dV \tag{12.1.31}$$

Tables of thermodynamic differentials from a condensed collection by P. W. Bridgman[10] give

$$(\partial P)_V \equiv -\left(\frac{\partial V}{\partial T}\right)_P$$

$$(\partial E)_V \equiv c_P\left(\frac{\partial V}{\partial P}\right)_T + T\left(\frac{\partial V}{\partial T}\right)_P^2 \quad \text{or}$$

$$(\partial P)_V = -\beta V$$

$$(\partial E)_V = -\kappa_T V c_P + T(\beta V)^2$$

We can take the ratio of the last two expressions to give

$$\left(\frac{\partial P}{\partial E}\right)_V = \frac{-\beta V}{-\kappa_T V c_P + T\beta^2 V^2} \qquad (12.1.32)$$

Also from thermodynamics, we have

$$c_P - c_v \equiv \frac{\beta^2 TV}{\kappa_T} \qquad (12.1.33)$$

Multiply relation (12.1.33) by $-V\kappa_T$:

$$-V\kappa_T c_P = -V\kappa_T c_v - V\frac{\kappa_T \beta^2 TV}{\kappa_T} = -V\kappa_T c_v - \beta^2 V^2 T$$

so that

$$-\kappa_T V c_P + \beta^2 V^2 T = -V\kappa_T c_v$$

Substituting in the denominator of relation (12.1.32) gives

$$\left(\frac{\partial P}{\partial E}\right)_V = \frac{\beta V}{V\kappa_T c_v} = \frac{\beta}{c_v \kappa_T} \qquad (12.1.34)$$

(This is also a standard thermodynamic result.) From relation (12.1.25) however, we have shown that $\gamma_G = V\left(\frac{\partial P}{\partial E}\right)_V$ so that we have

$$\boxed{\frac{\gamma_G}{V} = \frac{\beta}{c_v \kappa_T}} \qquad (12.1.35)$$

This is an important result that shows why it is usually a good approximation to take the Grüneisen coefficient as proportional to specific volume. The bulk thermal expansion coefficient, isothermal compression and specific heat are relatively slow functions of volume. Relation (12.1.31) allows us to calculate temperature along an isentrope, where $dS = 0$:

$$d \ln T = -\frac{\beta}{C_v \kappa_T} dV$$

We have

$$\ln\left(\frac{T}{T_0}\right) = -\int_{V_0}^{V} \frac{\beta}{C_v \kappa_T} dV$$

or

$$T = T_0 \exp\left(-\int_{V_0}^{V} \frac{\beta}{C_v \kappa_T} dV\right) = T_0 \exp\left(-\int_{V_0}^{V} \frac{\gamma_G(V)}{V} dV\right)$$

If we assume $\rho\gamma_G$ = constant then

$$\frac{\gamma_G(V)}{V} = \rho_0 \gamma_G(V_0) \equiv \rho_0 \gamma_0$$

and

$$T = T_0 e^{\rho_0 \gamma_0 (V_0 - V)}$$

Now consider forms of the Mie-Grüneisen EOS (Equation of State) used in hydrocodes.

12.1.2 Formulation for Hydrocode Use

Let's assume an arbitrary reference state P_0, E_0, $V_0 = 1/\rho_{0R}$. (The historic reason for the use of arbitrary reference states is the limit on size of numbers that could be handled by early computers.) From relation (12.1.22) we have

$$P_0 - P_k = \frac{\gamma_G}{V}(E_0 - E_k)$$

$$P - P_k - P_0 + P_k = \frac{\gamma_G}{V}(E - E_k - E_0 + E_k)$$

or

$$\boxed{P - P_0 = \frac{\gamma_G}{V}(E - E_0)} \qquad (12.1.36)$$

Also, from relation (12.1.24) we have

$$P - P_H = \frac{\gamma_G}{V}(E - E_H) \qquad (12.1.37)$$

But from the Hugoniot equations, we have

$$E_H - E_0 = \frac{1}{2}(P_H + P_0)(V_0 - V) \qquad (12.1.38)$$

From relation (12.1.36), $P >> P_0 \Rightarrow E >> E_0$, (or $P_H >> P_0 \Rightarrow E_H >> E_0$) so that we may write relation (11.1.38) approximately as

$$E_H \equiv \frac{1}{2}P_H(V_0 - V)$$

Now define $\mu \equiv \dfrac{\rho}{\rho_{0R}} - 1$, so that $(V_0 - V) = \mu V$ (We let $V_0 \equiv V_{0R}$) and

$$E_H \equiv \frac{1}{2}P_H \mu V \qquad (12.1.39)$$

Rewrite relation (12.1.37):

$$P = P_H + \frac{\gamma_G}{V}(E - E_H) = P_H\left(1 - \frac{\gamma_G E_H}{V P_H}\right) + \frac{\gamma_G}{V}E \qquad (12.1.40)$$

But from relation (12.1.39),

$$\frac{\gamma_G}{V}\frac{E_H}{P_H} \cong \frac{1}{2}P_H\mu V\frac{\gamma_G}{VP_H} = \frac{1}{2}\mu\gamma_G$$

Substituting in relation (12.1.40) gives

$$P \cong P_H\left(1-\frac{\gamma_G}{2}\mu\right)+\frac{\gamma_G}{V}E \qquad (P_H >> P_0) \qquad (12.1.41)$$

Now consider a cubic fit form for the Hugoniot:

$$D = c_0 + S_1 u + S_2\frac{u^2}{D} + S_3\frac{u^3}{D^2}$$

where u is the particle velocity and D is the shock velocity.

$$\boxed{\frac{c_0}{D} = 1 - S_1\left(\frac{u}{D}\right) - S_2\left(\frac{u}{D}\right)^2 - S_3\left(\frac{u}{D}\right)^3} \qquad (12.1.42)$$

The advantage of the form is that S_1, S_2, S_3 are dimensionless and thus the same for any system of units. We have

$$\frac{c_0}{D} = 1 - \sum_{n=1}^{3} S_n\left(\frac{u}{D}\right)^n$$

(Note that c_0 is the intercept for the reference state.) For $P_H >> P_0$, the Hugoniot equations give $P_H = \rho_0 Du$, which can be written

$$P_H = \frac{\rho_0 c_0^2\left(\frac{u}{D}\right)}{\left(\frac{c_0}{D}\right)^2} = \frac{\rho_0 c_0^2\left(\frac{u}{D}\right)}{\left[1-\sum_{n=1}^{3} S_n\left(\frac{u}{D}\right)^n\right]^2}$$

Substituting this in relation (12.1.41) gives

$$P \cong \frac{\rho_0 c_0^2 \left(\frac{u}{D}\right)}{\left[1-\sum_{n=1}^{3} S_n \left(\frac{u}{D}\right)^n\right]^2}\left(1-\gamma_G \frac{\mu}{2}\right)+\frac{\gamma_G}{V}E$$

which can be reduced further using conservation of mass ($u/D = \mu/(\mu+1)$) to give

$$P \cong \frac{\rho_0 c_0^2 \mu(\mu+1)(1-\mu\gamma_G/2)}{\left[(\mu+1)-\sum_{n=1}^{3} S_n \mu^n/(\mu+1)^{n-1}\right]^2}+\frac{\gamma_G}{V}E \qquad (12.1.43)$$

Now assume $\rho\gamma_G$ = constant and assume a model form:

$$\boxed{\gamma_G = \frac{\gamma_0 + a\mu}{(\mu+1)} = \gamma_0\left(\frac{V}{V_0}\right)+a\left(1-\frac{V}{V_0}\right)} \qquad (12.1.44)$$

Now

$$(\mu+1)(1-\mu\gamma_G/2) = \left[1+\left(1-\frac{\gamma_0}{2}\right)\mu-a\frac{\mu^2}{2}\right]$$

Substituting in relation (12.1.43) and expanding gives

$$\boxed{P = \frac{\rho_0 c_0^2 \mu\left[1+\left(1-\frac{\gamma_0}{2}\right)\mu-a\frac{\mu^2}{2}\right]}{\left[1-(S_1-1)\mu-S_2\frac{\mu^2}{(\mu+1)}-S_3\frac{\mu^3}{(\mu+1)^2}\right]^2}+(\gamma_0+a\mu)\rho_0 E} \qquad (12.1.45)$$

Now consider $\varepsilon \equiv \rho_{0R} E$. Common units for ε are MBar•cc/cc$_0$ .These units need some explanation. First define the quantities involved:

E = specific energy (energy/mass) = pressure • volume/mass
V = specific volume (volume/mass), so that
E/V = energy/volume = pressure.
γ_G, γ_0, a = dimensionless

$$\rho_{OR} = \frac{\text{mass}}{\text{volume}_0} \bullet \frac{\text{pressure}\bullet\text{volume}}{\text{mass}} = \frac{\text{pressure}\bullet\text{volume}}{\text{volume}_0}$$

thus, if pressures are expressed in MBar and volumes in cc, then the units of ε are MBar • cc/cc$_0$. A natural set of units for shock wave work is

$$P(\text{MBar}),\ u(\text{cm}/\mu\text{s}),\ \rho\left(\text{g/cm}^3\right),\ t(\mu\text{s})$$

We may rewrite relation (12.1.45) as

$$P = \frac{\rho_0 c_0^2 \mu\left[1+\left(1-\frac{\gamma_0}{2}\right)\mu - a\frac{\mu^2}{2}\right]}{\left[1-(S_1-1)\mu - S_2\frac{\mu^2}{(\mu+1)} - S_3\frac{\mu^3}{(\mu+1)^2}\right]^2} + (\gamma_0 + a\mu)\varepsilon \qquad (12.1.46)$$

This is the form generally found in hydrocodes. It incorporates the assumption $\rho\gamma_G$ = constant, using the model (12.1.44). For air with STP (Standard Temperature and Pressure) as the reference state, we have

$$S_1 = 1.81371 \times 10^{-3},\ S_2 = 1.07544 \times 10^{-2},\ S_3 = 1.51680 \times 10^{-2}$$
$$a = 2.92189,\ \gamma_0 = 6.52979,\ \rho_{OR} = 1.293 \times 10^{-3}\text{g/cm}^3,$$
$$c_0^2 = 6.22767 \times 10^{-3}\ (\text{cm}/\mu\text{s})^2$$

where the pressure is in megabars. The energy dependence of the Grüneisen EOS is so simple that it is trivial to invert and determine the specific internal energy corresponding to given initial pressure and density. This is useful for setting up problems corresponding to a given initial pressure and geometry.

Now consider transformation of derivatives in variables E, V to the variables ε, μ .

$$d\varepsilon = \rho_{OR}\, dE,\quad d\mu = \frac{1}{\rho_{OR}} d\rho$$

so that

$$\boxed{\begin{aligned}\left(\frac{\partial}{\partial E}\right)_V &= \rho_{0R}\left(\frac{\partial}{\partial \varepsilon}\right)_\mu \\ \left(\frac{\partial}{\partial \rho}\right)_E &= \frac{1}{\rho_{0R}}\left(\frac{\partial}{\partial \mu}\right)_\varepsilon\end{aligned}} \tag{12.1.47}$$

Now $\gamma_G = V\left(\frac{\partial P}{\partial E}\right)_V$ and $V = \frac{1}{\rho}$ so that $\gamma_G = \frac{\rho_{0R}}{\rho}\left(\frac{\partial P}{\partial \varepsilon}\right)_\mu$

but $\frac{\rho_{0R}}{\rho} = (\mu+1)^{-1}$ so that

$$\boxed{\gamma_G = (\mu+1)^{-1}\left(\frac{\partial P}{\partial \varepsilon}\right)_\mu} \tag{12.1.48}$$

Now consider the calculation of the sound speed.

$$dP = \left(\frac{\partial P}{\partial \rho}\right)_E d\rho + \left(\frac{\partial P}{\partial E}\right)_\rho dE$$

On an isentrope we have

$$dP_S = \left(\frac{\partial P}{\partial \rho}\right)_E d\rho_S + \left(\frac{\partial P}{\partial E}\right)_\rho dE_S$$

so that

$$\left(\frac{\partial P}{\partial \rho}\right)_S = \left(\frac{\partial P}{\partial \rho}\right)_E + \left(\frac{\partial P}{\partial E}\right)_\rho \left(\frac{\partial E}{\partial \rho}\right)_S \tag{12.1.49}$$

but $\gamma_G = V\left(\frac{\partial P}{\partial E}\right)_V = \frac{1}{\rho}\left(\frac{\partial P}{\partial E}\right)_\rho$ so that $\left(\frac{\partial P}{\partial E}\right)_\rho = \rho\gamma_G$, but on the isentrope

$$dE_S = -P dV_S = P\frac{d\rho_S}{\rho^2} \quad \text{and thus}$$

$$\left(\frac{\partial E}{\partial \rho}\right)_S = \frac{P}{\rho^2} \quad \text{so that} \quad \left(\frac{\partial P}{\partial E}\right)_\rho \left(\frac{\partial E}{\partial \rho}\right)_S = \frac{P\gamma_G}{\rho} \quad \text{We also have } c^2 = \left(\frac{\partial P}{\partial \rho}\right)_S .$$

Substitution in relation (12.1.49) gives

$$\boxed{c^2 = \left(\frac{\partial P}{\partial \rho}\right)_E + \frac{P}{\rho}\gamma_G} \tag{12.1.50}$$

Now transform to the variables ε, μ. Using relations (12.1.47) we can write

$$c^2 = \frac{1}{\rho_{OR}}\left(\frac{\partial P}{\partial \mu}\right)_\varepsilon + \frac{\gamma_G}{\rho}P$$

and since $\rho = (\mu+1)\rho_{OR}$ this becomes

$$\boxed{c^2 = \frac{1}{\rho_{OR}}\left(\frac{\partial P}{\partial \mu}\right)_\varepsilon + \frac{\gamma_G(\mu,\varepsilon)}{\rho_{OR}(\mu+1)}P(\mu,\varepsilon)} \tag{12.1.51}$$

Where $P(\mu,\varepsilon)$ is given by relation (12.1.46) and $\gamma_G(\mu,\varepsilon)$ is given by relation (12.1.48). Alternatively, we can write the result more explicitly as

$$c^2 = \frac{1}{\rho_{OR}}\left[\left(\frac{\partial P}{\partial \mu}\right)_\varepsilon + \frac{P}{(\mu+1)^2}\left(\frac{\partial P}{\partial \varepsilon}\right)_\mu\right]$$

For hydrocodes, both P and γ_G are of interest however, so relation (12.1.51) is a convenient choice. From relation (12.1.46) we have

$$\left(\frac{\partial P}{\partial \varepsilon}\right)_\mu = \gamma_0 + a\mu$$

which we see is consistent with relation (12.1.48), and our model (12.1.44) for γ_G . Now define

$$\xi(\mu) \equiv \left[1-(S_1-1)\mu - \frac{S_2\mu^2}{(\mu+1)} - \frac{S_3\mu^3}{(\mu+1)^2}\right]^2 \tag{12.1.52}$$

and

$$\chi(\mu) \equiv \mu\left[1+\left(1-\frac{\gamma_0}{2}\right)\mu-\frac{a\mu^2}{2}\right] \tag{12.1.53}$$

so that

$$P = \rho_{OR}\, c_0^2 \frac{\chi(\mu)}{\xi(\mu)} + (\gamma_0 + a\mu)\varepsilon \tag{12.1.54}$$

then

$$\left(\frac{\partial P}{\partial \mu}\right)_\varepsilon = \rho_{OR}\, c_0^2\left[\frac{\xi\chi'-\chi\xi'}{\xi^2}\right]+a\varepsilon \tag{12.1.55}$$

where the primes denote $\dfrac{d}{d\mu}$.

Specification of μ determines ξ and χ and their derivatives ξ' and χ' from relations (12.1.52) and (12.1.53) . Specification of ε then determines P from relation (12.1.54), and the derivative, relation (12.1.55) can be evaluated and substituted in relation (12.1.51) with γ_G from the model (12.1.44) and the pressure to give the sound speed.

It is instructive and of practical value to show how to reduce the form (12.1.42) to pressure versus mass velocity. We have

$$D = V_0\sqrt{\frac{P-P_0}{V_0-V}} \, , \quad u = \sqrt{(P-P_0)(V_0-V)}$$

therefore

$$\frac{u}{D} = 1 - \frac{V}{V_0} \, , \; \mu = \frac{\rho}{\rho_0} - 1$$

so that

$$\frac{u}{D} \equiv \xi = \left(\frac{\mu}{\mu+1}\right)$$

Assume that $P_0 = 0$, then

$$P = \rho_0 Du = \frac{\rho_0 c_0^2 (u/D)}{(c_0/D)^2} = \frac{\rho_0 c_0^2 \xi}{\left[1 - \sum_{n=1}^{3} s_n \xi^n\right]^2}$$

Assume ρ_0, c_0, S_1, S_2, S_3 are given. for a given compression ρ/ρ_0 (or μ) we calculate μ and then ξ

$$\left(\frac{c_0}{D}\right) = 1 - \sum_{n=1}^{3} S_n \xi^n$$

then gives D , and $P = \rho_0 Du$ gives pressure. By scanning ρ/ρ_0 (or μ) we can thus map out $P(\mu)$, $D(\mu)$. u is calculated from the expression for u/D .

12.1.3 Slater's Elastic Continuum Model

Slater's[11] calculation of γ_G is based on the Debye model for the phonon spectrum. A spherical Brillouin zone is assumed, which determines the maximum frequency.

$$\frac{dN}{d\upsilon} = 4\pi\upsilon^2 V\left(\frac{1}{c_l^3} + \frac{2}{c_t^3}\right)$$

where c_l, c_t = velocities for longitudinal and transverse phonons, and V is the volume of the crystal.

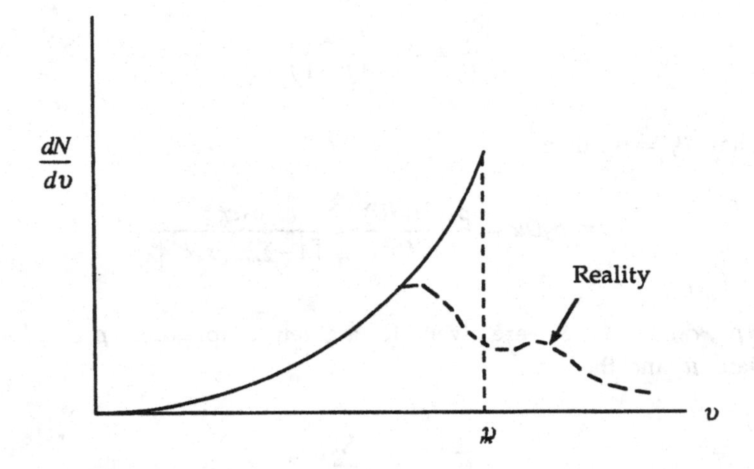

Fig. 12.1. Model phonon spectrum for Slater's Elastic Continuum model. The Debye model is assumed, giving a maximum frequency for the spectrum.

We assume $3N$ normal modes so that

$$\int_0^{\upsilon_m} dN = 3N \qquad \text{(total number of allowed frequencies)}$$

Substituting gives

$$3N = 4\pi V\left(\frac{1}{c_l^3}+\frac{2}{c_t^3}\right)\int_0^{\upsilon_m} \upsilon^2 d\upsilon = 4\pi V\left(\frac{1}{c_l^3}+\frac{2}{c_t^3}\right)\frac{\upsilon_m^3}{3}$$

so that

$$\frac{3N}{V} = \frac{4\pi}{3}\left(\frac{1}{c_l^3}+\frac{2}{c_t^3}\right)\upsilon_m^3$$

Solve for υ_m .

$$\frac{3N}{V}\cdot\frac{3}{4\pi}\left(\frac{1}{c_l^3}+\frac{2}{c_t^3}\right)^{-1} = v_m^3$$

and thus

$$v_m = \left[\frac{9N}{4\pi V}\left(\frac{1}{c_l^3}+\frac{2}{c_t^3}\right)^{-1}\right]^{\frac{1}{3}} \tag{12.1.56}$$

Now from elasticity theory we have for the longitudinal and transverse sound velocities:

$$c_l = \sqrt{\frac{3(1-\sigma)}{\kappa_T\rho(1+\sigma)}},\quad c_t = \sqrt{\frac{3(1-2\sigma)}{2\kappa_T\rho(1+\sigma)}}$$

where σ = Poisson's ratio, ρ = density, and κ_T = isothermal compressibility.

$$c_l^3 = \left[\frac{3(1-\sigma)}{\kappa_T\rho(1+\sigma)}\right]^{\frac{3}{2}},\quad c_t^3 = \left[\frac{3(1-2\sigma)}{2\kappa_T\rho(1+\sigma)}\right]^{\frac{3}{2}}$$

$$\left(\frac{1}{c_l^3}+\frac{2}{c_t^3}\right) = \left[\frac{\kappa_T\rho(1+\sigma)}{3(1-\sigma)}\right]^{\frac{3}{2}} + 2\left[\frac{2\kappa_T\rho(1+\sigma)}{3(1-2\sigma)}\right]^{\frac{3}{2}}$$
$$= (\kappa_T\rho)^{\frac{3}{2}}\left\{\left[\frac{(1+\sigma)}{3(1-\sigma)}\right]^{\frac{3}{2}} + 2\left[\frac{2(1+\sigma)}{3(1-2\sigma)}\right]^{\frac{3}{2}}\right\}$$

Define

$$f(\sigma) \equiv \left\{\left[\frac{(1+\sigma)}{3(1-\sigma)}\right]^{\frac{3}{2}} + 2\left[\frac{2(1+\sigma)}{3(1-2\sigma)}\right]^{\frac{3}{2}}\right\}$$

so that

$$\left(\frac{1}{c_l^3}+\frac{2}{c_t^3}\right) = (\kappa_T\rho)^{\frac{3}{2}}f(\sigma)$$

we then have

$$\upsilon_m = \left[\frac{9}{4\pi}\cdot\frac{N\rho}{(\rho\kappa_T)^{\frac{3}{2}}f(\sigma)}\right]^{\frac{1}{3}} = \left[\frac{9N}{4\pi f(\sigma)}\cdot\frac{\rho^{-\frac{1}{2}}}{\kappa_T^{\frac{3}{2}}}\right]^{\frac{1}{3}}$$

or

$$\upsilon_m = \left[\frac{9N}{4\pi f(\sigma)}\right]^{\frac{1}{3}}\cdot\frac{\rho^{-\frac{1}{6}}}{\kappa_T^{\frac{1}{2}}}$$

or finally,

$$\upsilon_m = \left[\frac{9N}{4\pi f(\sigma)}\right]^{\frac{1}{3}}\cdot\frac{V^{\frac{1}{6}}}{\kappa_T^{\frac{1}{2}}} \tag{12.1.57}$$

We assume the Grüneisen model: $-\dfrac{d\ln\upsilon_m}{d\ln V}$ is the same for all frequencies, so that

$$\gamma_G = -\frac{d\ln\upsilon_m}{d\ln V}$$

$$\ln\upsilon_m = \ln\left[\frac{9N}{4\pi f(\sigma)}\right]^{\frac{1}{3}} + \frac{1}{6}\ln V - \frac{1}{2}\ln\kappa_T$$

$$\frac{\partial\ln\upsilon_m}{\partial\ln V} = \frac{1}{6} - \frac{1}{2}\frac{\partial\ln\kappa_T}{\partial\ln V} - \frac{1}{3}\frac{\partial\ln f(\sigma)}{\partial\ln V}$$

so that

$$\gamma_G = -\frac{1}{6} + \frac{1}{2}\frac{\partial\ln\kappa_T}{\partial\ln V} + \frac{1}{3}\frac{\partial\ln f(\sigma)}{\partial\ln V} \tag{12.1.58}$$

At this point, Slater assumed Poisson's ratio to be independent of volume, so that the last term would drop out (not a very good assumption).

Now

$$\kappa_T = -\frac{1}{V}\left(\frac{\partial V}{\partial P}\right)_T$$

and

$$\ln \kappa_T = -\ln V + \ln\left[-\left(\frac{\partial V}{\partial P}\right)_T\right] = -\ln V - \ln\left[-\left(\frac{\partial P}{\partial V}\right)_T\right]$$

$$\frac{\partial \ln \kappa_T}{\partial \ln V} = -1 - \frac{1}{-\left(\frac{\partial P}{\partial V}\right)_T} \cdot V\frac{\partial}{\partial V}\left[-\left(\frac{\partial P}{\partial V}\right)_T\right] = -1 - V\frac{\left(\frac{\partial^2 P}{\partial V^2}\right)_T}{\left(\frac{\partial P}{\partial V}\right)_T}$$

we then have

$$\gamma_G = -\frac{1}{6} + \frac{1}{2}\left[-1 - V\frac{\left(\frac{\partial^2 P}{\partial V^2}\right)_T}{\left(\frac{\partial P}{\partial V}\right)_T}\right] = -\frac{1}{6} - \frac{1}{2} - \frac{V}{2}\frac{\left(\frac{\partial^2 P}{\partial V^2}\right)_T}{\left(\frac{\partial P}{\partial V}\right)_T}$$

or

$$\boxed{\gamma_G = -\frac{2}{3} - \frac{V}{2}\frac{\left(\frac{\partial^2 P}{\partial V^2}\right)_T}{\left(\frac{\partial P}{\partial V}\right)_T}} \qquad (12.1.59)$$

which is Slater's result.

Since $c_l = \sqrt{\frac{3(1-\sigma)}{\kappa_T \rho(1+\sigma)}}$ and $c_t = \sqrt{\frac{3(1-2\sigma)}{2\kappa_T \rho(1+\sigma)}}$, as σ increases, both c_l and c_t decrease, therefore $\left(\frac{1}{c_l^3} + \frac{2}{c_t^3}\right)$ increases, so that υ_m decreases. Now $\theta_D \equiv \frac{h\upsilon_m}{k}$ where h is Planck's constant, k is the Boltzmann constant, and θ_D is the Debye temperature, but as V increases σ increases ($\sigma \sim 1/3$ for solids, $1/2$ for liquids) so that θ_D decreases. γ_G will increase. Actually, c_l and c_t change differently with volume so that the spectrum must change. Grüneisen's assumption that $d\ln\upsilon_i/d\ln V$ is the same for all modes is thus not really justified. Note that for $\sigma = 1/2$ we have $c_t = 0$, and $c_l = 1/\sqrt{\rho\kappa_T}$ in the liquid, and for the solid, $\sigma \cong 1/3$ gives $c_t \sim c_l/2$.

12.1.4 Dugdale-MacDonald Model

Consider the potential energy ϕ of a crystal lattice to be some function of the Cartesian coordinates, $x_1, \ldots x_n$ of the atoms of the lattice. We have

$$P = -\frac{d\phi}{dV} = -\sum_i \frac{\partial\phi}{\partial x_i}\frac{dx_i}{dV} \tag{12.1.60}$$

Now

$$\frac{dP}{dV} = -\frac{d^2\phi}{dV^2} = -\sum_i \frac{d}{dV}\left(\frac{\partial\phi}{\partial x_i}\frac{dx_i}{dV}\right) = -\sum_i\left\{\frac{d}{dV}\left(\frac{\partial\phi}{\partial x_i}\right)\frac{dx_i}{dV} + \frac{\partial\phi}{\partial x_i}\frac{d}{dV}\left(\frac{dx_i}{dV}\right)\right\}$$

but

$$\frac{d}{dV}\left(\frac{\partial\phi}{\partial x_i}\right) = \frac{\partial}{\partial x_i}\left(\frac{d\phi}{dV}\right) = \frac{\partial}{\partial x_i}\left(\sum_j \frac{\partial\phi}{\partial x_j}\frac{dx_j}{dV}\right) = \sum_j\left(\frac{\partial^2\phi}{\partial x_i \partial x_j}\cdot\frac{dx_j}{dV} + \frac{\partial\phi}{\partial x_j}\cdot\frac{\partial}{\partial x_i}\frac{dx_j}{dV}\right)$$

Simplify notation: $$\frac{\partial^2\phi}{\partial x_i \partial x_j} \equiv \phi_{ij}$$

Now $$\frac{\partial}{\partial x_i}\frac{dx_j}{dV} = \frac{d}{dV}\left(\frac{\partial x_j}{\partial x_i}\right) = \begin{cases} \frac{d}{dV}(1) = 0 & for\, i = j \\ 0 & for\, i \neq j \end{cases}$$

so that $$\frac{\partial}{\partial x_i}\left(\frac{dx_j}{dV}\right) = 0 \text{ for all } i, j$$

We thus have $$\frac{d}{dV}\left(\frac{\partial\phi}{\partial x_i}\right) = \sum_j \phi_{ij}\frac{dx_j}{dV}$$ so that

$$\frac{dP}{dV} = -\sum_i\sum_j \phi_{ij}\frac{dx_j}{dV}\frac{dx_i}{dV} - \sum_i \frac{\partial\phi}{\partial x_i}\frac{d^2x_i}{dV^2} \tag{12.1.61}$$

We will consider cubic symmetry, so that $\dfrac{dx_i}{dV} = \dfrac{1}{3}\dfrac{x_i}{V}$ then we have

$$\frac{d^2x_i}{dV^2} = \frac{d}{dV}\left(\frac{1}{3}\frac{x_i}{V}\right) = \frac{1}{3V}\frac{dx_i}{dV} - \frac{1}{3}\frac{x_i}{V^2} = \frac{1}{3V}\left(\frac{1}{3}\frac{x_i}{V}\right) - \frac{1}{3V^2}x_i$$
$$= \left(\frac{1}{9} - \frac{1}{3}\right)\frac{x_i}{V^2} = -\frac{2}{9}\frac{x_i}{V^2}$$

Substituting gives

$$\frac{dP}{dV} = -\sum_{ij}\phi_{ij}\left(\frac{x_i}{3V}\right)\left(\frac{x_j}{3V}\right) - \sum_i \frac{\partial\phi}{\partial x_i}\left(-\frac{2}{9}\frac{x_i}{V^2}\right) = -\frac{1}{9V^2}\sum_{ij}\phi_{ij}x_ix_j + \frac{2}{9V^2}\sum_i\frac{\partial\phi}{\partial x_i}x_i$$

but

$$P = -\sum_i \frac{\partial\phi}{\partial x_i}\frac{dx_i}{dV} = -\sum_i\frac{\partial\phi}{\partial x_i}\left(\frac{x_i}{3V}\right) = -\frac{1}{3V}\sum_i\frac{\partial\phi}{\partial x_i}x_i$$

so that
$$\sum_i \frac{\partial\phi}{\partial x_i}x_i = -3PV$$

then substituting gives

$$\frac{dP}{dV} = -\frac{1}{9V^2}\sum_{ij}\phi_{ij}x_ix_j + \frac{2}{9V^2}(-3PV) = -\frac{1}{9V^2}\sum_{ij}\phi_{ij}x_ix_j - \frac{2}{3}\frac{P}{V}$$

so that
$$\frac{dP}{dV} + \frac{2}{3}\frac{P}{V} = -\frac{1}{9V^2}\sum_{ij}\phi_{ij}x_ix_j \tag{12.1.62}$$

which we rewrite as

$$-9V^2\left(P'+\frac{2P}{3V}\right) = \sum_{ij}\phi_{ij}x_ix_j$$

Differentiate this equation to obtain:

$$d\left[-9V^2\left(P'+\frac{2P}{3V}\right)\right] = \sum_{ij}d(\phi_{ij}x_ix_j)$$

but

$$d(\phi_{ij}x_ix_j) = x_ix_jd\varphi_{ij}+\varphi_{ij}(x_idx_j+x_jdx_i)$$

and

$$\frac{d(\phi_{ij}x_ix_j)}{d\ln V} = \frac{d\varphi_{ij}}{d\ln V}x_ix_j+\varphi_{ij}\left(x_i\frac{dx_j}{d\ln V}+x_j\frac{dx_i}{d\ln V}\right)$$

$$= \varphi_{ij}\frac{d\ln\varphi_{ij}}{d\ln V}x_ix_j+\varphi_{ij}V\left(x_i\frac{x_j}{3V}+\frac{x_i}{3V}x_j\right)$$

or

$$\frac{d(\phi_{ij}x_ix_j)}{d\ln V} = \varphi_{ij}x_ix_j\frac{d\ln\varphi_{ij}}{d\ln V}+\frac{2}{3}\phi_{ij}x_ix_j$$

so that

$$\frac{d}{d\ln V}\left[-9V^2\left(P'+\frac{2P}{3V}\right)\right] = \sum_{ij}\frac{d\ln\varphi_{ij}}{d\ln V}\varphi_{ij}x_ix_j+\frac{2}{3}\sum_{ij}\phi_{ij}x_ix_j \qquad (12.1.63)$$

Now the Mie-Grüneisen equation of state assumes that all the φ_{ij} change in the same proportion with volume (ie: $d\ln\varphi_{ij}/d\ln V$ are all equal and can be factored out of the sum.) The φ_{ij} are the interatomic force constants which govern the small amplitude thermal vibrations at a given volume. Consider a dynamically equivalent set of Harmonic Oscillators with force constants μ_α. The force constants are the roots of the secular equation[12]:

$$\begin{vmatrix} M\mu-\phi_{11} & -\phi_{12} & \cdot & \cdot \\ -\phi_{21} & M\mu-\phi_{22} & \cdot & \cdot \\ \cdot & & \cdot\ \cdot & \cdot \\ \cdot & & \cdot\ \cdot & M\mu-\phi_{nn} \end{vmatrix} = 0$$

where M is the atomic mass (assumed the same for all atoms, but this is not essential.) The roots vary as the φ_{ij} :

$$d\ln\mu_\alpha = d\ln\phi_{ij} \quad \text{for all } \alpha, \text{i, j}$$

For the harmonic oscillator, $d\ln\upsilon = \frac{1}{2}d\ln\mu_\alpha$ so that $2\frac{d\ln\upsilon}{d\ln V} = \frac{d\ln\phi_{ij}}{d\ln V}$

But we have $\gamma_G \equiv -\frac{d\ln\upsilon}{d\ln V}$ so that finally, we can write

$$\frac{d\ln\phi_{ij}}{d\ln V} = -2\gamma_G \qquad (12.1.64)$$

Substituting this result gives

$$\frac{d}{d\ln V}\left[-9V^2\left(P'+\frac{2P}{3V}\right)\right] = \left(\frac{2}{3}-2\gamma_G\right)\sum_{ij}\phi_{ij}x_ix_j$$

but we already know that

$$\sum_{ij}\phi_{ij}x_ix_j = -9V^2\left(P'+\frac{2P}{3V}\right)$$

so that

$$\frac{d}{d\ln V}\left[-9V^2\left(P'+\frac{2P}{3V}\right)\right] = \left(\frac{2}{3}-2\gamma_G\right)\left[-9V^2\left(P'+\frac{2P}{3V}\right)\right]$$

or

$$2\gamma_G = \frac{2}{3} - \frac{d\ln}{d\ln V}\left[-9V^2\left(P' + \frac{2P}{3V}\right)\right]$$

or

$$\gamma_G = \frac{1}{3} - \frac{1}{2}\frac{d\ln\left[-9V^2\left(P' + \frac{2}{3}\frac{P}{V}\right)\right]}{d\ln V} \tag{12.1.65}$$

$$\gamma_G = \frac{1}{3} - \frac{V}{2}\frac{d}{dV}\ln\left[-9V^2\left(P' + \frac{2P}{3V}\right)\right]$$

Further algebraic manipulation of this leads to

$$\gamma_G = \frac{1}{3} - \frac{V}{2}\frac{\left[3P''V^2 + 8P'V + 2P\right]}{\left[3P'V^2 + 2PV\right]} \tag{12.1.66}$$

Now consider $\omega \equiv PV^{\frac{2}{3}}$

$$\omega' \equiv \frac{d\omega}{dV} = P'V^{\frac{2}{3}} + \frac{2}{3}PV^{-\frac{1}{3}}$$

$$\omega'' \equiv P''V^{\frac{2}{3}} + \frac{2}{3}P'V^{-\frac{1}{3}} - \frac{2}{9}PV^{-\frac{4}{3}} + \frac{2}{3}P'V^{-\frac{1}{3}} = P''V^{\frac{2}{3}} + \frac{4}{3}P'V^{-\frac{1}{3}} - \frac{2}{9}PV^{-\frac{4}{3}}$$

$$\frac{\omega''}{\omega'} = \frac{P''V^{\frac{2}{3}} + \frac{4}{3}P'V^{-\frac{1}{3}} - \frac{2}{9}PV^{-\frac{4}{3}}}{P'V^{\frac{2}{3}} + \frac{2}{3}PV^{-\frac{1}{3}}}$$

Multiply numerator and denominator by $3V^{\frac{4}{3}}$ and expand to obtain

$$\frac{\omega''}{\omega'} = = \frac{3P''V^2 + 4P'V - \frac{2}{3}P}{3P'V^2 + 2PV} \tag{12.1.67}$$

Now pull $-2/3$ out of the second term in our expression for γ_G :

$$\gamma_G = \frac{1}{3} - \frac{2}{3} + \frac{2}{3} - \frac{V}{2}\frac{(3P''V^2 + 8P'V + 2P)}{(3P'V^2 + 2PV)}$$

Algebraic manipulation leads to

$$\gamma_G = -\frac{1}{3} - \frac{V}{2}\left[\frac{3P''V^2 + 4P'V - \frac{2}{3}P}{3P'V^2 + 2PV}\right]$$

but the term in brackets is just ω''/ω' so that

$$\gamma_G = -\frac{1}{3} - \frac{V\omega''}{2\omega'}$$

We thus have

$$\boxed{\gamma_G = -\frac{1}{3} - \frac{V}{2}\frac{\partial^2(PV^{2/3})/\partial V^2}{\partial(PV^{2/3})/\partial V}} \qquad (12.1.68)$$

This is the most frequently quoted form for the Dugdale-MacDonald formula, although it is not the original form they gave, which was

$$\gamma_G = -1 - \frac{V}{2}\left\{\frac{\partial^2 P/\partial V^2 - 10P/9V^2}{\partial P/\partial V + \frac{2}{3}P/V}\right\}$$

It can be shown with algebraic manipulation that the two expressions are equivalent. The Dugdale-MacDonald result was based on the assumption that the thermal expansion for a purely harmonic atomic potential is zero (incorrect). Although their result is incorrect it is found that in fact it does fit observed data better than the Slater model based on the elastic continuum, and thus it still finds frequent use.

12.2 Corrected Ideal Gas Equation of State

The ideal gas assumes the particles occupy negligible volume and do not interact. At high compressions this is not a very good assumption. We can correct for the particle volumes and interactions with a multiplicative function. The resulting EOS is useful for describing gases in applications such as a light-gas gun. The range of compressions for which it is valid is somewhat limited compared to models such as the ratio of polynomials EOS.

For the ideal gas we have $PV = RT$, $c_P - c_V = R$ and $c_V = \frac{dE}{dT}$ so that

$$E - E_0 = \int_{T_0}^{T} c_V \, dT$$

Assume c_V = constant and $E_0 = 0$ at $T_0 = 0$ so that $E = c_v T$.

Now $\frac{c_p - c_v}{c_v} = \frac{R}{c_v} = \gamma - 1$,where γ is the specific heat ratio, so that

$c_v = \frac{R}{(\gamma - 1)}$ and thus $E = \frac{RT}{(\gamma - 1)} = \frac{PV}{(\gamma - 1)}$ so that

$$\boxed{P = \frac{E}{V}(\gamma - 1)} \qquad (12.2.1)$$

This suggests the form

$$P = \frac{E}{V}(\gamma - 1) f(V, E)$$

where $f(V, E)$ is a correcting function. As before, define

$$\mu \equiv \frac{\rho}{\rho_{OR}} - 1, \quad \varepsilon \equiv \rho_{OR} E \quad \text{so} \quad (\mu + 1)\rho_{OR} = \rho = \frac{1}{V}, \quad E = \frac{\varepsilon}{\rho_{OR}}$$

therefore $\frac{E}{V} = (\mu + 1)\rho_{OR} \bullet \frac{\varepsilon}{\rho_{OR}} = (\mu + 1)\varepsilon$ and thus

$$\boxed{P = (\mu+1)\,\varepsilon(\gamma-1)f(\mu,\varepsilon)} \tag{12.2.2}$$

Choose the form

$$f(\mu,\varepsilon) = F_1(\mu)+F_2(\mu)\varepsilon+F_3(\mu)\varepsilon^2 \tag{12.2.3}$$

with

$$\left\{\begin{aligned} F_1(\mu) &= a_1+a_2\mu+a_3\mu^2 \\ F_2(\mu) &= a_4+a_5\mu+a_6\mu^2 \\ F_3(\mu) &= a_7+a_8\mu+a_9\mu^2 \end{aligned}\right\} \tag{12.2.4}$$

Now

$$\left(\frac{\partial P}{\partial \varepsilon}\right)_\mu = (\mu+1)(\gamma-1)f(\mu,\varepsilon)+(\mu+1)\varepsilon(\gamma-1)\left(\frac{\partial f}{\partial \varepsilon}\right)_\mu$$

but $\left(\frac{\partial f}{\partial \varepsilon}\right)_\mu = F_2(\mu)+2F_3(\mu)\varepsilon$ so that

$$\left(\frac{\partial P}{\partial \varepsilon}\right)_\mu = (\gamma-1)(\mu+1)\left[F_1(\mu)+F_2(\mu)\varepsilon+F_3(\mu)\varepsilon^2+F_2(\mu)\varepsilon+2F_3(\mu)\varepsilon^2\right]$$

or

$$\left(\frac{\partial P}{\partial \varepsilon}\right)_\mu = (\gamma-1)(\mu+1)\left[F_1(\mu)+2F_2(\mu)\varepsilon+3F_3(\mu)\varepsilon^2\right]$$

from relation (12.1.48) we have $\gamma_G = (\mu+1)^{-1}\left(\frac{\partial P}{\partial \varepsilon}\right)_\mu$ so that

$$\boxed{\gamma_G = (\gamma-1)\left[F_1(\mu)+2F_2(\mu)\varepsilon+3F_3(\mu)\varepsilon^2\right]} \tag{12.2.5}$$

Now

$$\left(\frac{\partial P}{\partial \mu}\right)_\varepsilon = \varepsilon(\gamma-1)f(\mu,\varepsilon)+(\mu+1)\varepsilon(\gamma-1)\left(\frac{\partial f}{\partial \mu}\right)_\varepsilon$$

$$\left(\frac{\partial f}{\partial \mu}\right)_\varepsilon = F_1'(\mu)+F_2'(\mu)\varepsilon+F_3'(\mu)\varepsilon^2 \tag{12.2.6}$$

where prime denotes $d/d\mu$.

$$\begin{Bmatrix} F_1'(\mu) = a_2+2a_3\mu \\ F_2'(\mu) = a_5+2a_6\mu \\ F_3'(\mu) = a_8+2a_9\mu \end{Bmatrix} \tag{12.2.7}$$

We thus have

$$\left(\frac{\partial P}{\partial \mu}\right)_\varepsilon = \varepsilon(\gamma-1)\left[(\mu+1)\left(\frac{\partial f}{\partial \mu}\right)_\varepsilon+f\right]$$

or

$$\left(\frac{\partial P}{\partial \mu}\right)_\varepsilon = \varepsilon(\gamma-1)\left[(\mu+1)\left\{F_1'(\mu)+F_2'(\mu)\varepsilon+F_3'(\mu)\varepsilon^2\right\}+F_1(\mu)+F_2(\mu)\varepsilon+F_3(\mu)\varepsilon^2\right]$$

so that

$$\left(\frac{\partial P}{\partial \mu}\right)_\varepsilon = \varepsilon(\gamma-1)[((\mu+1)F_1'(\mu)+F_1(\mu))+((\mu+1)F_2'(\mu)+F_2(\mu))\varepsilon$$
$$+((\mu+1)F_3'(\mu)+F_3(\mu))\varepsilon^2] \tag{12.2.8}$$

Specification of the density parameter μ determines the coefficients F_i and their derivatives from relations (12.2.4) and (12.2.7). The corresponding energy dependence is then determined for derivative (12.2.8) and for γ_G from relation (12.2.5). Specification of the energy parameter ε then determines the pressure from relations (12.2.2) - (12.2.3), γ_G from relation (12.2.5), derivative (12.2.8) and the sound speed from relation

(12.1.51). The specific internal energy corresponding to a given pressure and density merely involves solving a quadratic in ε.

12.3 Polynomial Equation of State

An empirical form often used with fluids is

$$\boxed{\begin{array}{l} P = c_0 + c_1\mu + c_2\mu^2 + c_3\mu^3 + \left(c_4 + c_5\mu + c_6\mu^2\right)\varepsilon \\ \text{with } c_2\mu^2,\ c_6\mu^2 \longrightarrow 0 \text{ for } \mu < 0 \text{ (tension)} \\ \mu = \dfrac{\rho}{\rho_0} - 1 = \dfrac{V_{0R}}{V} - 1,\ \rho = \rho_{0R}(\mu + 1) \end{array}} \qquad (12.3.1)$$

Now

$$\gamma_G = V\left(\frac{\partial P}{\partial E}\right)_V, \quad c^2 = \left(\frac{\partial P}{\partial \rho}\right)_E + \frac{P}{\rho}\gamma_G \qquad (12.3.2)$$

$$\varepsilon = \rho_{0R} E\ , \quad d\varepsilon = \rho_{0R}\, d\varepsilon\ , \quad d\mu = \frac{1}{\rho_{0R}} d\rho$$

so that we have

$$\left(\frac{\partial}{\partial E}\right)_V = \rho_{0R}\left(\frac{\partial}{\partial \varepsilon}\right)_\mu, \quad \left(\frac{\partial}{\partial \rho}\right)_E = \frac{1}{\rho_{0R}}\left(\frac{\partial}{\partial \mu}\right)_\varepsilon$$

$$\gamma_G = \frac{1}{\rho_{0R}(\mu + 1)} \cdot \rho_{0R}\left(\frac{\partial P}{\partial \varepsilon}\right)_\mu = (\mu + 1)^{-1}\left(\frac{\partial P}{\partial \varepsilon}\right)_\mu$$

$$\left(\frac{\partial P}{\partial \varepsilon}\right)_\mu = \left(c_4 + c_5\mu + c_6\mu^2\right)$$

thus

$$\gamma_G = \frac{(c_4 + c_5\mu + c_6\mu^2)}{(\mu+1)} \tag{12.3.3}$$

$$\left(\frac{\partial P}{\partial \rho}\right)_E = \frac{1}{\rho_{OR}}\left[c_1 + 2c_2\mu + 3c_3\mu^2 + (c_5 + 2c_6\mu)\varepsilon\right]$$

and $1/\rho = 1/\rho_{OR}(\mu+1)$ so that

$$c^2 = \frac{1}{\rho_{OR}}\left[c_1 + 2c_2\mu + 3c_3\mu^2 + (c_5 + 2c_6\mu)\varepsilon\right] + \frac{P(\mu,\varepsilon)}{\rho_{OR}(\mu+1)} \cdot \gamma_G(\mu) \tag{12.3.4}$$

12.4 Ratio of Polynomials Equation of State

This is another empirical form often used with fluids. As before,

$$\mu = \frac{\rho}{\rho_{OR}} - 1, \quad \varepsilon = \rho_{OR}E$$

The form used is:

$$P(\mu,\varepsilon) = (1+\mu)\frac{F_1(\mu) + F_2(\mu)\varepsilon + F_3(\mu)\varepsilon^2 + F_4(\mu)\varepsilon^3}{F_5(\mu) + F_6(\mu)\varepsilon + F_7(\mu)\varepsilon^2} \tag{12.4.1}$$

$$\text{where} \quad F_i(\mu) = \sum_{j=1}^{4} a_l \mu^{j-1} \quad \text{with } l = j + 4(i-1)$$

The functions $F_i(\mu)$ are all cubic:

$$F_1(\mu) = a_1 + a_2\mu + a_3\mu^2 + a_4\mu^3$$
$$F_2(\mu) = a_5 + a_6\mu + a_7\mu^2 + a_8\mu^3$$
$$\cdot \quad \cdot \quad \cdot \quad \cdot \quad \cdot$$
$$\cdot \quad \cdot \quad \cdot \quad \cdot \quad \cdot$$
$$F_7(\mu) = a_{25} + a_{26}\mu + a_{27}\mu^2 + a_{28}\mu^3$$

The form is thus the ratio of a cubic in energy over a quadratic in energy, with coefficients that are cubic in the volume parameter. The form thus has 28 adjustable coefficients and is more flexible than the polynomial or Mie-Grüneisen forms. In order to use the form, one must specify both the coefficients and the reference density. In the process of fitting, one can constrain the fit to give agreement with observed data such as the sound speed, Grüneisen parameter, etc. Now consider the form of the pressure, relation (12.4.1). We define:

$$\psi(\mu,\varepsilon) \equiv F_1(\mu)+F_2(\mu)\varepsilon+F_3(\mu)\varepsilon^2+F_4(\mu)\varepsilon^3 \tag{12.4.2}$$

$$\omega(\mu,\varepsilon) \equiv F_5(\mu)+F_6(\mu)\varepsilon+F_7(\mu)\varepsilon^2 \tag{12.4.3}$$

We have

$$P(\mu,\varepsilon) = (1+\mu)\,\psi(\mu,\varepsilon)/\omega(\mu,\varepsilon) \tag{12.4.4}$$

$$\left(\frac{\partial\psi}{\partial\mu}\right)_\varepsilon = F_1'(\mu) + F_2'(\mu)\varepsilon+F_3'(\mu)\varepsilon^2+F_4'(\mu)\varepsilon^3 \tag{12.4.5}$$

$$\left(\frac{\partial\omega}{\partial\mu}\right)_\varepsilon = F_5'(\mu) + F_6'(\mu)\varepsilon+F_7'(\mu)\varepsilon^2 \tag{12.4.6}$$

Where the primes denote $d/d\mu$.

$$\left(\frac{\partial\psi}{\partial\varepsilon}\right)_\mu = F_2(\mu)+2F_3(\mu)\varepsilon+3F_4(\mu)\varepsilon^2 \tag{12.4.7}$$

$$\left(\frac{\partial\omega}{\partial\varepsilon}\right)_\mu = F_6(\mu)+2F_7(\mu)\varepsilon \tag{12.4.8}$$

We also have

$$\left\{\begin{aligned} F_1'(\mu) &= a_2+2a_3\mu+3a_4\mu^2 \\ F_2'(\mu) &= a_6+2a_7\mu+3a_8\mu^2 \\ F_3'(\mu) &= a_{10}+2a_{11}\mu+3a_{12}\mu^2 \\ F_4'(\mu) &= a_{14}+2a_{15}\mu+3a_{16}\mu^2 \\ F_5'(\mu) &= a_{18}+2a_{19}\mu+3a_{20}\mu^2 \\ F_6'(\mu) &= a_{22}+2a_{23}\mu+3a_{24}\mu^2 \\ F_7'(\mu) &= a_{26}+2a_{27}\mu+3a_{28}\mu^2 \end{aligned}\right\} \tag{12.4.9}$$

Now differentiating relation (12.4.4) gives

$$\left(\frac{\partial P}{\partial \mu}\right)_\varepsilon = \frac{\psi}{\omega} + (1+\mu)\left[\frac{\omega\left(\frac{\partial \psi}{\partial \mu}\right)_\varepsilon - \psi\left(\frac{\partial \omega}{\partial \mu}\right)_\varepsilon}{\omega^2}\right] \tag{12.4.10}$$

and

$$\left(\frac{\partial P}{\partial \varepsilon}\right)_\mu = (1+\mu)\left[\frac{\omega\left(\frac{\partial \psi}{\partial \varepsilon}\right)_\mu - \psi\left(\frac{\partial \omega}{\partial \varepsilon}\right)_\mu}{\omega^2}\right] \tag{12.4.11}$$

For a given μ, relation (12.4.1) determine the $F_i(\mu)$ coefficients and relations (12.4.9) determine their derivatives. The forms ψ and ω in relations (12.4.2) and (12.4.3) are thus determined as functions of energy, as are the forms of the derivatives (12.4.5) - (12.4.8). Specification of ε then determines the functions ψ and ω from relations (12.4.2) and (12.4.3). The pressure is then determined from relation (12.4.4). Derivatives (12.4.5) - (12.4.8) can then be evaluated and substituted to determine the derivatives (12.4.10) and (12.4.11). The derivative (12.4.11) may then be substituted in relation (12.1.48) to give the Grüneisen coefficient. Finally, the derivative (12.4.10) may be used in relation (12.1.51) along with the pressure and Grüneisen coefficient to give the sound speed corresponding to the state (μ, ε). In practice, the above steps are readily compressed into a few lines of code for efficient calculation, but this is the calculational sequence used. The analytically calculated sound speed is used in controlling the time step in hydrocodes. It is useful to invert the EOS to give the specific internal energy corresponding to a given pressure and density, as when setting up initial conditions for a hydrocode problem. The pressure relation (12.4.1) can be arranged as a cubic in the energy: $A\varepsilon^3 + B\varepsilon^2 + C\varepsilon + D = 0$ where

$$A = (1+\mu)F_4(\mu), \quad B = (1+\mu)F_3(\mu) - PF_7(\mu)$$
$$C = (1+\mu)F_2(\mu) - PF_6(\mu), \quad D = (1+\mu)F_1(\mu) - PF_5(\mu)$$

or in standard form: $\varepsilon^3 + A_0\varepsilon^2 + B_0\varepsilon + C_0 = 0$ where

$$A_0 = B/A, \quad B_0 = C/A, \quad C_0 = D/A$$

so that

$$A_0 = \frac{(1+\mu)F_3(\mu) - PF_7(\mu)}{(1+\mu)F_4(\mu)}$$

$$B_0 = \frac{(1+\mu)F_2(\mu) - PF_6(\mu)}{(1+\mu)F_4(\mu)}$$

$$C_0 = \frac{(1+\mu)F_1(\mu) - PF_5(\mu)}{(1+\mu)F_4(\mu)}$$

Once μ and P are specified, the real coefficients A_0, B_0, C_0 can be calculated. An analytic solution of the arbitrary cubic with real coefficients is known. In general, for this application, the solution usually has one real root and two complex roots which are conjugates of each other. We want the real root. The method of solution will be deferred to an appendix. As an example, we will give a fit for hydrogen that is appropriate for use with light-gas guns. For hydrogen we have

$$\rho_{STP} = 8.99972 \times 10^{-5} \text{ g/cm}^3 \quad \text{or}$$
$$V_{STP} = 11111.457 \text{ cm}^3/\text{g}$$
$$P_{STP} = 1.0133 \times 10^{-6} \text{ MBar}$$

The fit coefficients are as follows:

$$a_1 = 1.00350 \times 10^{-13},\ a_2 = -1.2397 \times 10^{-16},\ a_3 = 3.4151 \times 10^{-20},$$
$$a_4 = -4.3955 \times 10^{-24},\ a_5 = 4.01508 \times 10^{-1},\ a_6 = 4.00661 \times 10^{-1},$$
$$a_7 = -6.38420 \times 10^{-4},\ a_8 = 1.71374 \times 10^{-7},\ a_9 = -8.79622 \times 10^{3},$$
$$a_{10} = -8.38224 \times 10^{3},\ a_{11} = 3.72627 \times 10^{2},\ a_{12} = 2.50517 \times 10^{-1},$$
$$a_{13} = 1.21948 \times 10^{8},\ a_{14} = 1.38452 \times 10^{8},\ a_{15} = 1.67264 \times 10^{7},$$
$$a_{16} = -5.84837 \times 10^{2},\ a_{17} = 1.00000 \times 10^{0},\ a_{18} = 9.98509 \times 10^{-1},$$
$$a_{19} = -1.20279 \times 10^{-3},\ a_{20} = 2.64023 \times 10^{-7},\ a_{21} = -2.34174 \times 10^{4},$$
$$a_{22} = -2.27354 \times 10^{4},\ a_{23} = 6.02167 \times 10^{2},\ a_{24} = -1.46940 \times 10^{-1},$$
$$a_{25} = 5.34381 \times 10^{8},\ a_{26} = 6.02326 \times 10^{8},\ a_{27} = 6.94633 \times 10^{7},$$
$$a_{28} = -1.09953 \times 10^{4},\ \rho_{0R} = 8.18 \times 10^{-5} \text{ (g/cm}^3\text{)}$$

The fit gives

$$E_{STP} = 0.02798 \text{ MBar} \cdot \text{cc/g} = 2.798 \text{ kJ/g}$$

The fit should above should be good to about 10% for the range $10^{-2} \le V/V_{STP} \le 10^1$, $10^2 \le T(K) \le 2.5 \times 10^3$. A considerable additional range is covered with the same tolerance, but the regions are not easily described by a simple block of values for the temperature and volume.

12.5 Tillotson Equation of State

In 1962 Tillotson developed a simplified EOS package as part of project DEFENDER[13] sponsored by the Advanced Research Projects Agency (ARPA). It is an analytic representation joining Hugoniot data and Thomas-Fermi calculations. There are only two phases: solid (condensed) and vapor (expanded). Melting is ignored. There is no tension model. It is only applicable for processes such as impacts. A situation involving external energy input during compression cannot be handled. The solid is described by a modified Mie-Grüneisen EOS which transitions smoothly to ideal gas behavior at large energy where Thomas-Fermi results are adequate. The form for the gas is chosen to make

$$P, \left(\frac{\partial P}{\partial E}\right)_V, \left(\frac{\partial P}{\partial V}\right)_E$$

and hence c^2 and γ_G continuous across the phase line. At large volumes the form approaches the ideal gas behavior. At low energy, where a shock is not strong enough for anything but elastic behavior, a region is provided to allow expanded states without evaporation. Otherwise, the phase line bordering the condensed and expanded phases is taken to be normal volume. For the condensed phase, the form is

$$\boxed{P = \left[a + \frac{b}{\left(\frac{E}{E_0 \eta^2} + 1\right)}\right]\frac{E}{V} + A\mu + B\mu^2} \qquad (12.5.1)$$

$$\text{where } V = \frac{1}{\rho},\ \eta = \frac{\rho}{\rho_0} = \frac{V_0}{V},\ \mu = \eta - 1$$

The constants a, b, A are determined by fits to EOS data, while E_0 and B are adjusted for best overall fit to the $P(E,V)$ data. Calculate the Grüneisen parameter:

$$\left(\frac{\partial P}{\partial E}\right)_V = \left[\frac{-b/E_0\eta^2}{\left(\frac{E}{E_0\eta^2}+1\right)^2}\right]\frac{E}{V} + \left[a + \frac{b}{\left(\frac{E}{E_0\eta^2}+1\right)}\right]\frac{1}{V}$$

so that

$$\gamma_G = V\left(\frac{\partial P}{\partial E}\right)_V = a + \frac{b}{\left(\frac{E}{E_0\eta^2}+1\right)}\left\{1 - \frac{E/E_0\eta^2}{\left(\frac{E}{E_0\eta^2}+1\right)}\right\}$$

or

$$\boxed{\gamma_G = a + \frac{b}{\left(\frac{E}{E_0\eta^2}+1\right)^2}} \tag{12.5.2}$$

Now calculate $(\partial P/\partial V)_E$:

$$\frac{d\mu}{dV} = -\frac{V_0}{V^2},\ \frac{d\eta}{dV} = -\frac{V_0}{V^2}$$

$$\left(\frac{\partial P}{\partial V}\right)_E = \left[-b\left(\frac{E}{E_0\eta^2}+1\right)^{-2}\left(-\frac{2E}{E_0\eta^3}\frac{d\eta}{dV}\right)\right]\frac{E}{V} + \left[a + \frac{b}{\left(\frac{E}{E_0\eta^2}+1\right)}\right]\left(-\frac{E}{V^2}\right) + A\frac{d\mu}{dV} + 2B\mu\frac{d\mu}{dV}$$

Reducing this gives

$$\left(\frac{\partial P}{\partial V}\right)_E = -\frac{aE}{V^2} - (A+2B\mu)\frac{\eta}{V} - \frac{\frac{bE}{V^2}\left[\frac{3E}{E_0\eta^2}+1\right]}{\left(\frac{E}{E_0\eta^2}+1\right)^2}$$

Sound speed can be calculated from

$$c^2 = \left(\frac{\partial P}{\partial \rho}\right)_E + \frac{P}{\rho}\gamma_G, \quad V = \frac{1}{\rho}, \quad dV = -\frac{1}{\rho^2}d\rho$$

$$\left(\frac{\partial}{\partial \rho}\right)_E = -V^2\left(\frac{\partial}{\partial V}\right)_E$$

so that

$$\boxed{c^2 = -V^2\left(\frac{\partial P}{\partial V}\right)_E + P\gamma_G V} \tag{12.5.3}$$

Now for $E >> E_0$ the second term in the expression for γ_G becomes negligible so that

$$\gamma_G_{E>>E_0} \equiv a$$

The Thomas-Fermi result for γ_G thus determines the parameter a. For $E = 0$ (where $P = 0, \rho = \rho_0, \mu = 0, \eta = 1$) we have $\gamma_G = a+b$ so that γ_G at normal conditions along with the above determined value for a determines the parameter b. Calculation of the bulk modulus gives

$$M \equiv -V\left(\frac{\partial P}{\partial V}\right)_E = \frac{aE}{V} + (A+2B\mu)\eta + \frac{bE}{V}\frac{\left(\frac{3E}{E_0\eta^2}+1\right)}{\left(\frac{E}{E_0\eta^2}+1\right)^2}$$

At $E = 0$, where $\eta = 1, \mu = 0$ we have $M_0 = A$, so that we see A is the bulk modulus at normal conditions. In order to determine values for A we can use low pressure Hugoniot data: $M_0 = A = \rho_0 c_0^2$

A proper choice for c_0 determines A so that the initial slope of the Hugoniot is consistent with the slope of the low pressure shock data (usually c_0 is the intercept from the linear D, u fit, $D = c_0 + su$ unless a phase transition causes this not to extrapolate properly. In such cases use the adiabatic sound speed.)

For the expanded phase the EOS form is

$$P = \frac{aE}{V} + \left[\frac{b}{\left(\frac{E}{E_0 \eta^2} + 1\right)} \left(\frac{E}{V}\right) + A\mu e^{-\beta\left(\frac{V}{V_0} - 1\right)} \right] e^{-\alpha\left(\frac{V}{V_0} - 1\right)^2} \tag{12.5.4}$$

The first term is ideal gas behavior. At large volumes the exponentials force the ideal gas behavior to dominate. The form is chosen to make

$$P, \left(\frac{\partial P}{\partial E}\right)_V, \left(\frac{\partial P}{\partial V}\right)_E$$

continuous across the $V = V_0$ phase line. (eg; at $V/V_0 = 1$, we have $\eta = 1$, $\mu = 0$ and

$$P = \frac{aE}{V_0} + \frac{b}{\left(\frac{E}{E_0} + 1\right)} \frac{E}{V_0} = \left[a + \frac{b}{\left(\frac{E}{E_0} + 1\right)} \right] \frac{E}{V_0}$$

(which matches the condensed phase there.)

A serious problem with the Tillotson EOS is that the above form for the expanded phase is not monotonic at low energies and for V slightly greater than V_0, the term involving the bulk modulus can cause the pressure to go strongly negative. In addition, the sound speed calculation blows up with an imaginary result. In general, one must explore with the model to determine the minimum volume for which the form may be used.

Now consider the derivatives of the pressure:

$$\varphi(E,V) \equiv \left[\frac{b}{\left(\frac{E}{E_0\eta^2}+1\right)}\left(\frac{E}{V}\right)+A\mu e^{-\beta\left(\frac{V}{V_0}-1\right)}\right]$$

$$\psi(V) \equiv e^{-\alpha\left(\frac{V}{V_0}-1\right)^2}$$

so that

$$P = \frac{aE}{V} + \varphi(E,V)\,\psi(V)$$

and

$$\left(\frac{\partial P}{\partial E}\right)_V = \frac{a}{V} + \psi(V)\left(\frac{\partial \varphi}{\partial E}\right)_V$$

$$\left(\frac{\partial \varphi}{\partial E}\right)_V = \left(\frac{\partial}{\partial E}\right)_V\left[\frac{b}{\left(\frac{E}{E_0\eta^2}+1\right)}\left(\frac{E}{V}\right)\right] = \frac{b}{\left(\frac{E}{E_0\eta^2}+1\right)^2}\frac{1}{V}$$

$$\left(\frac{\partial P}{\partial E}\right)_V = \frac{a}{V} + \left[\frac{b}{\left(\frac{E}{E_0\eta^2}+1\right)^2}\left(\frac{1}{V}\right)\right] e^{-\alpha\left(\frac{V}{V_0}-1\right)^2}$$

since $\gamma_G = V\left(\frac{\partial P}{\partial E}\right)_V$ we have

$$\boxed{\gamma_G = a + \frac{b}{\left(\frac{E}{E_0\eta^2}+1\right)^2} e^{-\alpha\left(\frac{V}{V_0}-1\right)^2}} \qquad (12.5.5)$$

(Note that $\gamma_G = \gamma_G(E,V)$ here.) For either $E \gg E_0$ or $V \gg V_0$ we have $\gamma_G \longrightarrow a$. Now at $V = V_0$ or $\eta = 1$ we have

$$\gamma_G = a + \frac{b}{\left(\frac{E}{E_0}+1\right)^2}$$

which matches the result for the condensed phase. Now consider $\left(\frac{\partial P}{\partial V}\right)_E$:
As before,

$$P = \frac{aE}{V} + \varphi(E,V)\,\psi(V)$$

$$\left(\frac{\partial P}{\partial V}\right)_E = -\frac{aE}{V^2} + \left(\frac{\partial \varphi}{\partial V}\right)_E \psi(V) + \varphi(E,V)\frac{d\psi}{dV}$$

but

$$\frac{d\psi}{dV} = \frac{2\alpha}{V_0}\left(1-\frac{V}{V_0}\right)e^{-\alpha\left(\frac{V}{V_0}-1\right)^2} = \frac{2\alpha}{V_0}\left(1-\frac{V}{V_0}\right)\psi(V)$$

$$\left(\frac{\partial \varphi}{\partial V}\right)_E = -\frac{b}{\left(\frac{E}{E_0\eta^2}+1\right)}\left(\frac{E}{V}\right)\left[\frac{1}{V}+\frac{2V_0 E/E_0 V^2\eta^3}{\left(\frac{E}{E_0\eta^2}+1\right)}\right] - A\left(\frac{V_0}{V^2}+\frac{\beta\mu}{V_0}\right)e^{-\beta\left(\frac{V}{V_0}-1\right)}$$

At $V = V_0,\ \eta = 1,\ \mu = 0$ we have

$$\left(\frac{\partial P}{\partial V}\right)_E = -\frac{aE}{V_0^2} - \frac{A}{V_0} - \frac{bE}{V_0^2}\frac{\left(\frac{3E}{E_0}+1\right)}{\left(\frac{E}{E_0}+1\right)^2}$$

which is continuous with the value for the condensed phase. Since the derivatives are continuous across the boundary, the sound speed is also.

In Tillotson's original report the region of expanded elastic states was bounded by an energy E_s and a volume $V_s > V_0$. For $E < E_s$ and $V_0 < V < V_s$ the condensed state form was used. E_s is determined from thermodynamic data and is equivalent to the total specific internal energy at the boiling point. In practice it was necessary to add a fraction of the vaporization energy E_v to this to give reasonable behavior.

$$E_s' = E_s + \varepsilon E_v, \quad \text{with } \varepsilon \leq 0.2$$

E was then compared with E_s' rather than E_s in determining the phase boundary. More recently, a modified version of the Tillotson EOS has been used at Los Alamos National Laboratory (LANL) in the LASOIL code. An interpolation region between the two phases has been added for $E_s < E < E_s'$.

$$P = \frac{P_e(E-E_s)+P_c(E_s'-E)}{(E_s'-E_s)}$$

is used for the pressure. The derivatives are similarly interpolated. P_e, P_c are the expanded and condensed state pressures, respectively.

$$\left(\frac{\partial P}{\partial V}\right)_E = \frac{(\partial P_e/\partial V)_E(E-E_s)+(\partial P_c/\partial V)_E(E_s'-E)}{(E_s'-E_s)}$$

so that this derivative interpolates the same way. Now

$$\left(\frac{\partial P}{\partial E}\right)_V = \frac{\left[(\partial P_e/\partial E)_V(E-E_s)+P_e+(\partial P_c/\partial E)_V(E_s'-E)-P_c\right]}{(E_s'-E_s)}$$

or

$$\left(\frac{\partial P}{\partial E}\right)_V = \frac{\left[(\partial P_e/\partial E)_V(E-E_s)+(\partial P_c/\partial E)_V(E_s'-E)\right]}{(E_s'-E_s)}+\frac{P_e-P_c}{E_s'-E_s}$$

so that the energy derivative has an additional term. Inclusion of this term however would produce a discontinuity in the derivative at either end, so it is ignored. LANL has avoided the problem of negative pressures in the expanded state by substituting other forms in the relevant region at the price of creating discontinuities. At high energies, the usual assumption that the Grüneisen parameter is a function of volume only, cannot be justified. The form developed here for both condensed and expanded states depends on both volume and energy. The EOS is adequate up to about 1000 Mbar (±10%) and is claimed to be accurate to ≈ 5% below 5 Mbar. An approximate set of fit parameters for a limited number of elements is given in tables 12.5.1a and 12.5.1b. No provision has been made for the negative pressure problem. An interested user must provide his own solution on a material by material basis.

Table 12.5.1a Tillotson EOS Fit Parameters

	ρ_0 (g/cm^3)	a	E_0 (MBar•cc/g)	b	A (MBar)
Al	2.7	0.5	0.050	1.63	0.752
Be	1.845	0.55	0.175	0.62	1.173
Cd	8.65	0.5	0.045	1.7	0.473
Cu	8.9	0.5	0.325	1.5	1.39
Fe	7.86	0.5	0.095	1.5	1.279
Pb	11.34	0.4	0.015	2.37	0.4664
Mo	10.2	0.5	0.045	1.02	2.713
Ni	8.86	0.5	0.09	1.33	1.912
Th	11.68	0.4	0.025	0.86	0.531
Ti	4.51	0.5	0.070	0.60	1.03
W	19.17	0.5	0.225	1.04	3.08

Table 12.5.1b Tillotson EOS Fit Parameters

	E_s (MBar•cc/g)	E_s' (MBar•cc/g)	α	β	B (MBar)
Al	0.0276	0.141	5	5	0.65
Be	0.0631	0.419	5	5	0.55
Cd	0.0023	0.0143	10	10	0.45
Cu	0.0138	0.03	5	5	1.10
Fe	0.0142	0.0845	5	5	1.05
Pb	0.00203	0.0114	13	15	0.15
Mo	0.0162	0.0674	5	5	1.65
Ni	0.0168	0.0811	5	5	1.50
Th	0.00742	0.0334	9	0.88	0.5
Ti	0.0236	0.122	5	5	0.5
W	0.010	0.582	10	10	2.5

12.6 Appy Equation of State

An older analytic EOS called the Appy equation of state is still in use. It explicitly treats compressed and expanded states separately. It is based on the Mie-Grüneisen EOS and is primarily useful for solids. As before, we define

$$\mu \equiv \frac{\rho}{\rho_0} - 1, \quad \varepsilon \equiv \rho_0 E$$

then for $\mu \geq 0$ (compressed states) we have

$$P = P_H\left(1 - \frac{1}{2}\gamma_0 \mu\right) + \gamma_0 \varepsilon$$

where

$$P_H = \frac{\rho_0 a^2 \mu(\mu+1)}{\left[(b-1)\mu - 1\right]^2}, \quad \gamma_0 = \frac{g_0 + g_1 \mu}{(1 + g_2 \mu + g_3 \mu^2)}$$

$$g_0 = 2b - 1, \quad g_1 = \frac{3}{4} g_0^2 - \frac{1}{3} g_0 - 0.6388889$$

$$g_2 = g_0, \quad g_3 = g_0^2 - b^2$$

and for $\mu < 0$ (expanded states)

substitute
$$P = \frac{\rho_0 a^2 \mu(|\mu| + 1)}{\left[(b-1)\mu - 1\right]^2}$$

Example: Mylar, $\rho_0 = 1.39$ g/cm^3, $a^2 = 4.84 \times 10^{10}$ (cm/s)^2, $b = 1.6313$

Now from relation (12.1.48) we have

$$\gamma_G = (\mu + 1)^{-1}\left(\frac{\partial P}{\partial \varepsilon}\right)_\mu$$

$$\left(\frac{\partial P}{\partial \varepsilon}\right)_\mu = \gamma_0(\mu), \quad \gamma_G = \frac{\gamma_0(\mu)}{(\mu + 1)}$$

Now therefore,

$$\gamma_G = \frac{g_0 + g_1 \mu}{(1 + g_2 \mu + g_3 \mu^2)(\mu + 1)}$$

From relation (12.1.51) with $\rho_{0R} = \rho_0$, we have

$$c^2 = \frac{1}{\rho_0}\left(\frac{\partial P}{\partial \mu}\right)_\varepsilon + \frac{\gamma_G P}{\rho_0 (\mu + 1)}$$

Now define

$$\begin{aligned}
\varphi_1(\mu) &\equiv 1 + g_2 \mu + g_3 \mu^2 \\
\varphi_2(\mu) &\equiv \left[(b-1)\mu - 1\right]^2 \\
\varphi_3(\mu) &\equiv a^2\left(\frac{2\mu + 1}{2}\right)(2 - \gamma_0 \mu)\varphi_2^{-1}(\mu) \\
\varphi_4(\mu) &\equiv -\frac{a^2 \mu(\mu + 1)}{2}(\gamma_0 + \mu \gamma_0{}')\varphi_2^{-1}(\mu) \\
\varphi_5(\mu) &\equiv -\frac{a^2 \mu(\mu + 1)}{2}(2 - \gamma_0 \mu)\varphi_2^{-2}(\mu)\varphi_2{}'(\mu)
\end{aligned}$$

and

$$\omega(\mu) \equiv a^2 \mu(\mu + 1)\left(1 - \frac{\gamma_0}{2}\mu\right)\varphi_2^{-1}(\mu)\,, \quad \omega'(\mu) = \varphi_3(\mu) + \varphi_4(\mu) + \varphi_5(\mu)$$

then

$$P = \frac{\rho_0 a^2 \mu(\mu + 1)}{\varphi_1(\mu)}$$

and

$$\left(\frac{\partial P}{\partial \mu}\right)_\varepsilon = \rho_0\, \omega'(\mu) + \gamma_0{}'(\mu)\varepsilon$$

and

$$\begin{aligned}
\gamma_0(\mu) &= (g_0 - g_1 \mu)/\varphi_1(\mu) \\
\gamma_0{}'(\mu) &= -\frac{g_1}{\varphi_1(\mu)} - \frac{(g_0 - g_1 \mu)\varphi_1{}'(\mu)}{\varphi_1^2(\mu)}
\end{aligned}$$

The sound speed can be calculated by substitution in the expression for c^2 .

12.7 PUFF Equation of State

The PUFF code developed at the Air Force Weapons Laboratory uses an EOS that assumes either solid or vapor phases. Compressed states are treated as solid, while expanded states are assumed to be vapor. A modified polynomial equation of state is used for the compressed states and an unusual exponential form for expanded states. In the solid phase, the stress is assumed to be composed of the hydrostatic pressure and a stress deviator. For compressed states, the following describes the hydrostat. We define

$$\mu \equiv \frac{\rho}{\rho_0} - 1, \quad \varepsilon \equiv \rho_0 E, \quad \eta \equiv \frac{\rho}{\rho_0}$$

For $\mu \geq 0$ (compressed states) we have

$$P = a_1 \mu + a_2 \mu^2 + a_3 \mu^3 + b_1 (\mu + 1) \varepsilon$$

for $\mu < 0$ (expanded states) we have

$$P = \left\{ \left[\varepsilon - (1 - \eta^n) \varepsilon_s \right] e^{D\eta^{\frac{1}{3}}} \right\} c_1 \eta$$

where

$$D = \ln\left(\frac{b_1}{c_1}\right), \quad n = \frac{a_1}{b_1 \varepsilon_s}$$

For OTWR (an aerospace material):

$$a_1 = 0.0942 \text{ MBar}, a_2 = 0.0458 \text{ MBar}, a_3 = 0, b_1 = 0.64,$$
$$c_1 = 0.25, \varepsilon_s = 0.191 \text{ MBar}, \rho_0 = 1.7 \text{ g/cm}^3$$

Now for $\mu \geq 0$ we have

$$\gamma_G = (\mu + 1)^{-1} \left(\frac{\partial P}{\partial \varepsilon}\right)_\mu = b_1$$

$$c^2 = \frac{1}{\rho_0}\left(\frac{\partial P}{\partial \mu}\right)_\varepsilon + \frac{\gamma_G P}{\rho_0(\mu+1)}$$

$$\left(\frac{\partial P}{\partial \mu}\right)_\varepsilon = a_1 + 2a_2\mu + 3a_3\mu^2 + b_1\varepsilon$$

$$c^2 = \frac{1}{\rho_0}\left\{(a_1 + 2a_2\mu + 3a_3\mu^2 + b_1\varepsilon) + b_1\left[\frac{a_1\mu + a_2\mu^2 + a_3\mu^3 + b_1(\mu+1)\varepsilon}{(\mu+1)}\right]\right\}$$

and for $\mu < 0$ we have

$$\left(\frac{\partial P}{\partial \varepsilon}\right)_\mu = \left(\frac{\partial P}{\partial \varepsilon}\right)_\eta = c_1 \eta e^{D\eta^{\frac{1}{3}}}$$

so that

$$\gamma_G = \frac{1}{(\mu+1)}\left(\frac{\partial P}{\partial \varepsilon}\right)_\mu = c_1 e^{D\eta^{\frac{1}{3}}}$$

$$c^2 = \frac{1}{\rho_0}\left(\frac{\partial P}{\partial \mu}\right)_\varepsilon + \frac{\gamma_G}{\rho_0 \eta}P$$

$$\left(\frac{\partial P}{\partial \mu}\right)_\varepsilon = \left(\frac{\partial P}{\partial \eta}\right)_\varepsilon$$

After some algebra, we obtain

$$\left(\frac{\partial P}{\partial \eta}\right)_\varepsilon = \left(\frac{1}{\eta} + \frac{1}{3}D\eta^{-\frac{2}{3}}\right)P + nc_1\eta^n\varepsilon_s e^{D\eta^{\frac{1}{3}}}$$

so that

$$c^2 = \frac{1}{\rho_0}\left\{\left(\frac{1}{\eta} + \frac{1}{3}D\eta^{-\frac{2}{3}}\right)P + nc_1\eta^n\varepsilon_s e^{D\eta^{\frac{1}{3}}}\right\} + \frac{P}{\rho_0\eta}c_1 e^{D\eta^{\frac{1}{3}}}$$

12.8 Multi-branched Analytic Equation of State

For some applications, the equation of state must span enormous ranges of density and temperature, and include a realistic representation of the mixed phase regions. The multi-branched analytic equation of state[14] was developed to handle this while retaining the speed of an analytic calculation. The following example was developed for lithium in the range ($10^{-6}\,\mathrm{g/cm^3} \le \rho \le 2\,\mathrm{g/cm^3}$, $454\,\mathrm{K} \le T \le 3.7\times 10^{8}\,\mathrm{K}$). Accurate tables of P and E were constructed for this range. The range is too wide for any single EOS model, so three separate models were used. A soft-sphere model was used for the low pressure liquid and vapor region, an ionization-equilibrium model for the low density ionized gas, and a pseudopotential model for the liquid metal up to high density and temperature. For use in a hydrocode however, analytic representations of these models are much more efficient. A series of analytic forms was constructed to represent the various regions with smooth join between the forms. ρ, E space was divided into three main regions (see figure 12.2). Region 3 containing the two-phase (LV) region is most complex and is subdivided further. The melting point $T = 454\,\mathrm{K}$, $\rho_0 = 0.518\,\mathrm{g/cm^3}$ is taken to be the zero of energy. We first consider the pressure models.

Region 1: $(\rho \ge \rho_0)$

For region 1, the Grüneisen model is adequate and the Hugoniot is convenient as a reference rather than the cold compression curve. The form is

$$P = P_H\left(1 - \frac{\gamma_0}{2}\eta\right) + \gamma_0\rho_0 E \qquad (12.8.1)$$

where

$$P_H = \frac{\rho_0 c_0^2 \eta}{(1 - s\eta)^2}\,, \quad \eta = 1 - \frac{\rho_0}{\rho} \qquad (12.8.2)$$

and c_0 = sound speed at the melting point and s is the slope of the D, u (shock velocity - particle velocity) line, and γ_0 is the Grüneisen parameter. The values used for Lithium are:

$$\rho_0 = 0.518 \text{ g/cm}^3, \; \gamma_0 = 0.9, \; c_0 = 4.5 \text{ km/s}, \; s = 1.133$$

Region 2 $(\rho \le \rho_0, \, E > E_c)$

This region is above the critical point. The pressure should join smoothly to the ideal gas at $\rho > 0$ and to the Grüneisen model at $\rho = \rho_0$. For an ideal gas we can write

$$P = \xi(E - E_c)\rho \tag{12.8.3}$$

(For no internal degrees of freedom, $\xi = 2/3$) A suitable form is

$$P = \xi(E - E_c)\rho + \left[\gamma_0 \rho_0 E - \xi(E - E_c)\rho_0\right](\rho/\rho_0)^k \tag{12.8.4}$$

At $\rho = \rho_0, \eta = 0$ so that $P_H = 0$ and the pressure in region 1 is $P = \gamma_0 \rho_0 E$. Also $(\rho/\rho_0)^k = 1$ so that the terms in $(E - E_c)$ above cancel, giving $P = \gamma_0 \rho_0 E$ in region 2, giving smooth join. As $\rho \to 0$, the factor $(\rho/\rho_0)^k \to 0$ and the pressure in region 2 becomes the ideal gas law. The factor k giving best overall fit to the tabular data is 1.5.

The ideal gas form above is good for both low temperature (atomic Li) and high temperature (fully ionized Li), but in the partially ionized region there are strong deviations from the form, caused by ionization of the $2S^1$ and $1S^2$ electron shells. A multiplication factor $g(E)$ was devised to represent the deviation. The revised form is

$$P(\rho, E) = \xi(E - E_c)\rho g(E) + \left[\gamma_0 \rho_0 E - \xi(E - E_c)\rho_0 g(E)\right](\rho/\rho_0)^k \tag{12.8.5}$$

with

$$g(E) = 1 - 0.7e^{-2(y-1.233)^2} - 0.7e^{-0.5(y-4.468)^2} \tag{12.8.6}$$

where $y = \ln(E/E_c)$

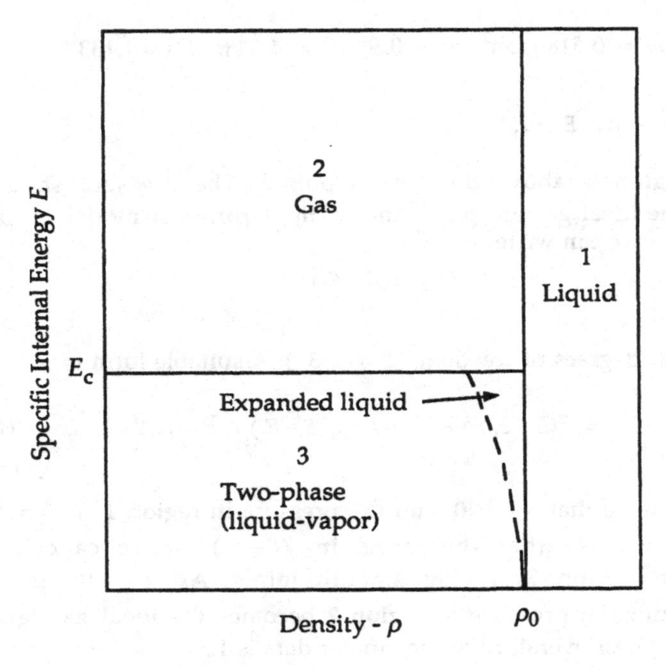

Fig. 12.2. The energy-density plane is divided into three regions for constructing an analytic equation of state.

As before $P = \gamma_0 \rho_0 E$ in both regions 1 and 2 at $\rho = \rho_0$ and

$$P = \xi(E - E_c)\rho g(E) \quad \text{as} \quad \rho \to 0 \tag{12.8.7}$$

Figure 12.3 shows the results for $\rho = 10^{-6}\text{g/cm}^3$.

Region 3 $(\rho \le \rho_0,\ E < E_c)$

For slightly expanded liquid $(\rho \le \rho_0)$ an appropriate simple form is

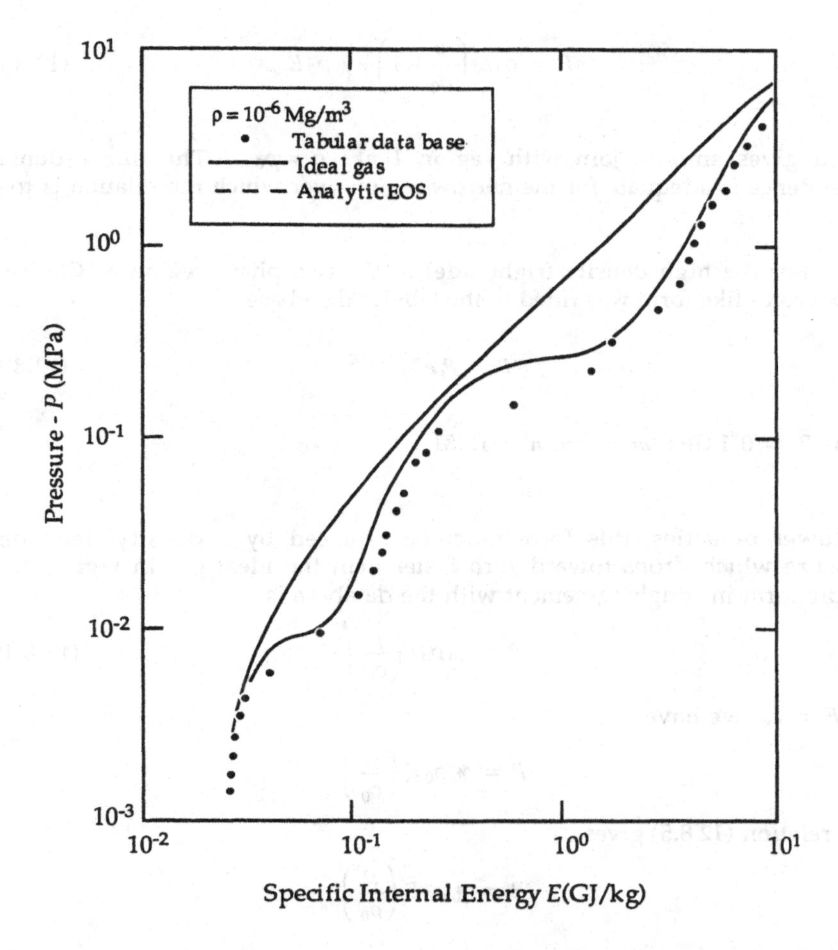

Fig. 12.3. A comparison of pressure versus specific internal energy at $\rho = 10^{-6}$ Mg/m^3 (which is numerically equal to g/cm^3) for the ideal gas model, the tabular base values and the analytic fit to the data base. The pressure is in MegaPascals (1 Bar = 10^5 MPa).

$$P = \rho_0 c_0^2 \left(\frac{\rho}{\rho_0} - 1 \right) + \gamma_0 \rho_0 E \qquad (12.8.8)$$

which gives smooth join with region 1 at $\rho = \rho_0$. The linear density dependence is adequate for the narrow region over which the relation is to be used.

For the high density (right-side) of the two-phase region a Clausius-Clapeyron - like form was fitted to the tabular data base:

$$P = P_c e^{m - nE_c/E} \qquad (12.8.9)$$

with $P_c = 0.1$ GPa, $m = 1.4$, $n = 1.481$

At lower densities, this form must be replaced by a density dependent pressure which drops toward zero faster than the ideal gas in region 2. A simple form in rough agreement with the database is

$$P = \gamma_0 \rho_0 E \left(\frac{\rho}{\rho_0} \right)^k \qquad (12.8.10)$$

At $E = E_c$ we have

$$P = \gamma_0 \rho_0 E_c \left(\frac{\rho}{\rho_0} \right)^k$$

and relation (12.8.5) gives

$$P = \gamma_0 \rho_0 E_c \left(\frac{\rho}{\rho_0} \right)^k$$

so that smooth join with region 2 is achieved at $E = E_c$. An interpolation region is needed between regions 2 and 3 to avoid discontinuities in thermodynamic derivatives. Accordingly, a region $E_1 \le E \le E_c$ is provided to give smooth join.

Below the interpolation region $(\rho < \rho_0,\ E \leq E_1 \leq E_c)$:

Figure 12.4 illustrates the calculation in the liquid-vapor transition region. The pressure is treated as constant through the phase transition (actually there is some slope.) In the transition we have

$$P_{51} = P_c e^{m - nEc/E} \tag{12.8.11}$$

which is independent of density. The numerical subscripts in the figure are labels used in the coding.

At the vapor side of the transition region, the iso-energy curve is described by

$$P_{52} = \gamma_0 \rho_0 E \left(\frac{\rho}{\rho_0} \right)^k \tag{12.8.12}$$

The boundary of the transition is determined by calculating both P_{51} and P_{52}. The smaller result is taken as P_5. The intersection of the two curves represents the boundary. At the liquid side of the transition region, the iso-energy curve is represented as

$$P_4 = \rho_0 c_0^2 \left(\frac{\rho}{\rho_0} - 1 \right) + \gamma_0 \rho_0 E \tag{12.8.13}$$

P_4 is calculated and compared with P_5. The larger value is accepted as the pressure P. The intersection represents the boundary.

In the interpolation region $(\rho < \rho_0,\ E_1 \leq E \leq E_c)$

In the transition region we use the form

$$P_{511} = P_c e^{m - nE_c/E} \tag{12.8.14}$$

At the vapor side, the iso-energy curve is represented by

$$P_{512} = \gamma_0 \rho_0 E \left(\frac{\rho}{\rho_0} \right)^k \tag{12.8.15}$$

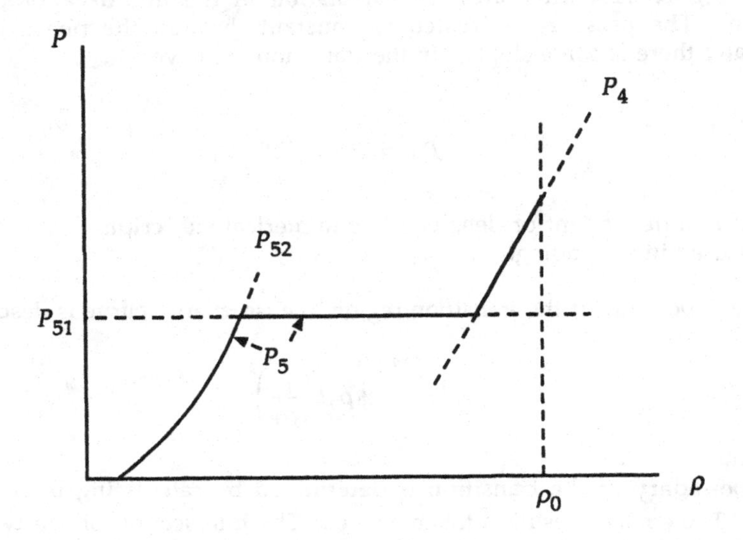

Fig. 12.4. Details of the calculation of the liquid-vapor transition region. The horizontal portion is in the coexistence region.

Both P_{511} and P_{512} are calculated and P_{51} is taken as the smaller of the two. At the liquid side the form

$$P_{41} = \rho_0 c_0^2 \left(\frac{\rho}{\rho_0} - 1 \right) + \gamma_0 \rho_0 E_1 \tag{12.8.16}$$

is calculated, as well as

$$P_2 = \gamma_0 \rho_0 E_c \left(\frac{\rho}{\rho_0} \right)^k \tag{12.8.17}$$

The larger of P_{41} and P_{51} is used to interpolate with P_2. (eg; if P_{51} is larger, interpolate between P_{51} and P_2 with the energy E between E_1 and E_c)

$$P = P_{51} + \frac{(P_2 - P_{51})}{(E_c - E_1)}(E - E_1)$$

Sound speed calculations

Sound speed can be calculated thermodynamically from

$$c^2 = \left(\frac{\partial P}{\partial \rho}\right)_E + \frac{P}{\rho}\gamma_G$$

where $\gamma_G = \frac{1}{\rho}\left(\frac{\partial P}{\partial E}\right)_\rho$ is the Grüneisen parameter, thus

$$c^2 = \left(\frac{\partial P}{\partial \rho}\right)_E + \frac{P}{\rho^2}\left(\frac{\partial P}{\partial E}\right)_\rho \qquad (12.8.18)$$

We will calculate the various forms for the derivatives corresponding to the various analytic pressure forms.

Region 1 $(\rho \geq \rho_0)$

$$P = P_H\left(1 - \frac{\gamma_0}{2}\eta\right) + \gamma_0 \rho_0 E$$

$$P_H = \frac{\rho_0 c_0^2 \eta}{(1 - s\eta)^2}, \quad \eta = 1 - \frac{\rho_0}{\rho}$$

$$\frac{d\eta}{d\rho} = \frac{\rho_0}{\rho^2}, \quad \left(\frac{\partial P_H}{\partial \rho}\right)_E = \left(\frac{dP_H}{d\eta}\right)_E \frac{d\eta}{d\rho} = \frac{\rho_0}{\rho^2}\left(\frac{dP_H}{d\eta}\right)_E$$

$$\left(\frac{dP}{d\eta}\right)_E = \left(\frac{dP_H}{d\eta}\right)_E\left(1 - \frac{\gamma_0}{2}\eta\right) - \frac{\gamma_0}{2}P_H$$

$$\frac{dP_H}{d\eta} = \frac{\rho_0 c_0^2}{(1-s\eta)^2} + \frac{2\rho_0 c_0^2 s\eta}{(1-s\eta)^3}$$

$$= \rho_0 c_0^2 \left[\frac{1}{(1-s\eta)^2} + \frac{2s\eta}{(1-s\eta)^3} \right]$$

so that

$$\left(\frac{dP}{d\eta}\right)_E = \rho_0 c_0^2 \frac{\left[1-\gamma_0\eta+2s\eta\left(1-\frac{\gamma_0}{2}\eta\right)/(1-s\eta)\right]}{(1-s\eta)^2} \tag{12.8.19}$$

then

$$\left(\frac{\partial P}{\partial \rho}\right)_E = \frac{\rho_0}{\rho^2}\left(\frac{\partial P}{\partial \eta}\right)_E \tag{12.8.20}$$

$$\left(\frac{\partial P}{\partial E}\right)_\rho = \gamma_0 \rho_0 \tag{12.8.21}$$

Region 2 $(\rho \leq \rho_0,\ E_c < E)$

$$P(\rho,E) = \xi(E-E_c)\rho g(E) + \left[\gamma_0\rho_0 E - \xi(E-E_c)\rho_0 g(E)\right]\left(\frac{\rho}{\rho_0}\right)^k$$

with

$$g(E) = 1-0.7e^{-2(y-1.233)^2}-0.7e^{-0.5(y-4.468)^2}$$

$$\text{where} \quad y = \ln\left(\frac{E}{E_c}\right)$$

$$\left(\frac{\partial P}{\partial E}\right)_\rho = \xi\rho g(E) + \xi(E-E_c)\rho g'(E) +$$

$$\left[\gamma_0\rho_0 - \xi\rho_0 g(E) - \xi(E-E_c)\rho_0 g'(E)\right]\left(\frac{\rho}{\rho_0}\right)^k \tag{12.8.22}$$

Now

$$g'(E) = \frac{dg}{dy}\frac{dy}{dE}, \quad \frac{dy}{dE} = \frac{1}{(E/E_c)}\left(\frac{1}{E_c}\right) = \frac{1}{E}$$

$$\frac{dg}{dy} = 2.8(y-1.233)e^{-2(y-1.233)^2} + 0.7(y-4.468)e^{-0.5(y-4.468)^2}$$

so that

$$g'(E) = \frac{1}{E}[2.8(y-1.233)e^{-2(y-1.233)^2} + 0.7(y-4.468)e^{-0.5(y-4.468)^2}] \tag{12.8.23}$$

Now therefore

$$\left(\frac{\partial P}{\partial \rho}\right)_E = \xi(E-E_c)g(E) + \left[\gamma_0 \rho_0 E - \xi(E-E_c)\rho_0 g(E)\right]k\left(\frac{\rho}{\rho_0}\right)^k \frac{1}{\rho} \tag{12.8.24}$$

Region 3. ($\rho \leq \rho_0$, $E < E_c$)

First consider the form

$$P = \rho_0 c_0^2\left(\frac{\rho}{\rho_0} - 1\right) + \gamma_0 \rho_0 E$$

We have

$$\left(\frac{\partial P}{\partial E}\right)_\rho = \gamma_0 \rho_0, \quad \left(\frac{\partial P}{\partial \rho}\right)_E = c_0^2 \tag{12.8.25}$$

Now consider the form

$$P = P_c e^{m - nE_c/E}$$

We have

$$\left(\frac{\partial P}{\partial \rho}\right)_E = 0 , \quad \left(\frac{\partial P}{\partial E}\right)_\rho = \frac{nE_c}{E^2} P_c e^{m-nE_c/E} = \frac{nE_c}{E^2} P \tag{12.8.26}$$

Finally, consider the form

$$P = \gamma_0 \rho_0 E \left(\frac{\rho}{\rho_0}\right)^k$$

We have

$$\left(\frac{\partial P}{\partial E}\right)_\rho = \gamma_0 \rho_0 \left(\frac{\rho}{\rho_0}\right)^k = \frac{P}{E} \tag{12.8.27}$$

$$\left(\frac{\partial P}{\partial \rho}\right)_E = \gamma_0 \rho_0 E \left(\frac{\rho}{\rho_0}\right)^{k-1} \bullet \frac{k}{\rho_0} = \gamma_0 \rho_0 E k \left(\frac{\rho}{\rho_0}\right)^k \frac{1}{\rho} = \frac{k}{\rho} P \tag{12.8.28}$$

Temperature Calculation

Temperature calculation is provided for radiation transport calculation

Region 1:

$$T = \frac{(E - E_0)}{3R} \tag{12.8.29}$$

where R is the gas constant and

$$E_0 = E_{00} + \gamma_0 E_{00} \eta + \frac{1}{2}\left(c_0^2 + \gamma_0^2 E_{00}\right)\eta^2 + \frac{1}{6}\left(4sc_0^2 + \gamma_0^3 E_{00}\right)\eta^3 \tag{12.8.30}$$

with

$$E_{00} = -3RT_0 \qquad (T_0 = 454\ \mathrm{K})$$

Region 2:

$$T = T_G + \left(\frac{\rho}{\rho_0}\right)^{0.2} (T_s - T_G) \tag{12.8.31}$$

where

$$T_G = \left(\frac{1}{2}+\frac{7}{2}\frac{E_c}{E}\right)\frac{(E-E_c)}{6R} \tag{12.8.32}$$

$$T_s = \frac{E}{3R}+T_0 \tag{12.8.33}$$

Relation (12.8.31) deals approximately with the transition from an atomic gas (1 particle/atom) at low temperature to a fully ionized gas (4 particles/atom) at high temperature. The temperature in region 2 may not be strictly a function of specific internal energy and density however, since at high temperatures and low density, the bulk of the energy will be in the form of radiation fields. The hydrocode supplies density and the combined energy of matter and fields

$$E_T = E + \frac{4\sigma}{c\rho}T^4$$

where σ is the Stephan-Boltzmann constant. Equation (12.8.31) is then used to begin an iterative solution for the temperature. The radiation pressure

$$\frac{1}{3}\left(\frac{4\sigma}{c}\right)T^4$$

is added to the material pressure to give the total pressure supplied to the hydrocode.

Region 3:

$$T = \left(\frac{\rho}{\rho_0}\right)^{0.2}\left(\frac{E}{3R}+T_0\right) \tag{12.8.34}$$

It can be seen from relations (12.8.31 - 12.8.33) that the temperature is continuous across the border between regions 2 and 3, but not between regions 1 and 3 or 1 and 2. These temperatures are adequate for radiation transport, but are not thermodynamically consistent with the rest of the model.

Results:

Figure 12.5 shows various iso-energy curves for the model in the P, ρ plane. It can be seen that the curve $E = E_c$ is the curve passing through the critical point. The rollover for $\rho > \rho_0$ shows the transition to the Grüneisen model. Again, it is evident that the pressure is treated as independent of density in the two-phase region (an approximation). A listing in BASIC of

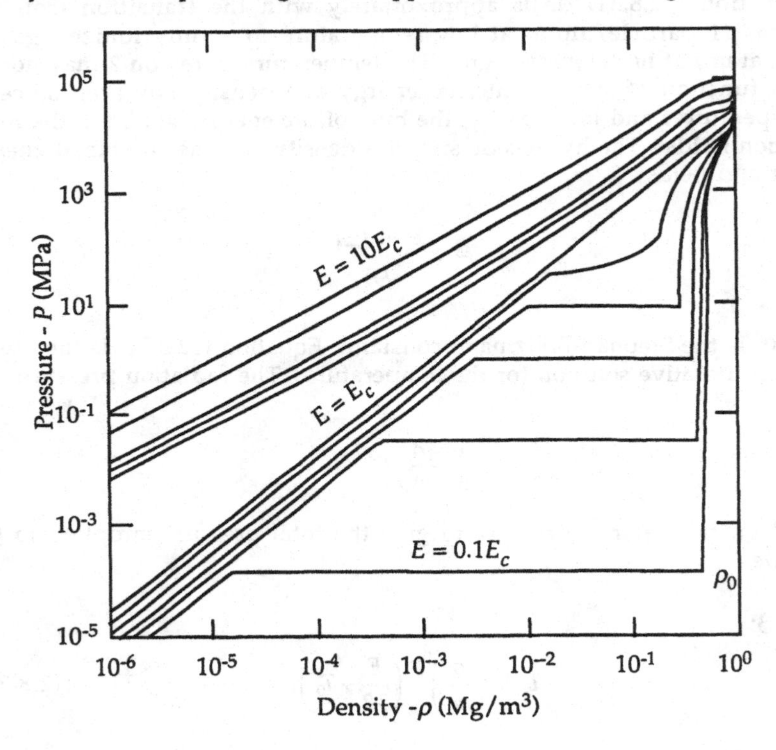

Fig. 12.5. Pressure as a function of density for various values of specific internal energy, computed from the analytical EOS for Li. The pressure is in MegaPascals (1 Bar = 10^5 MPa). The density is in Mg/m^3 (which is numerically equal to g/cm^3).

the coding for the EOS is given in the appendices, with the constants for Lithium. A simple driver is included to allow the model to be explored. The subroutine for g(E) and g'(E) begins at line 1600. The EOS model begins at line 1000.

In summary, the constants used for Li are: $\rho_0 = 0.518$ g/cm^3, $T_0 = 454$ K, $P_c = 0.1$ GPa, $E_c = 23.03$ MJ/kg, $E_1 = 14$ MJ/kg, $\gamma_0 = 0.9$, $c_0 = 4.5$ km/s, $s = 1.133$, $m = 1.4$, $n = 1.481$, $k = 1.5$, $\xi = 2/3$, and the gas constant $R = 1.198$ kJ/kg•K

12.9 Modified Soft-Sphere Equation of State

The modified soft-sphere equation of state was developed to describe liquid metals. Young[15] made modifications to the soft sphere theory of Hoover, et al[16]. The model is based on Monte Carlo calculations for particles interacting with pair potentials of the form $\varphi(r) = \varepsilon(\sigma/r)^n$, where r is the interparticle distance, σ is the sphere diameter, and $4 \le n \le 12$. It uses a fairly simple configurational heat capacity term

$$\frac{\delta\Phi}{NkT} = \frac{1}{6}(n+4)\rho^{n/9}(\varepsilon/kT)^{1/3}$$

The kinetic energy $\frac{3}{2}NkT$ and a static lattice term $C_n\rho^{n/3}$ are subtracted from the Monte Carlo energy. $\rho = N\sigma^3/\sqrt{2}V$ and C_n is the fcc Madelung constant for the potential, which may be approximated by

$$C_n \cong 6.0+\frac{6.669}{(n-3)} - 1.043(n-4)^{0.389}\exp\left[-0.156(n-4)\right]$$

Young adapted the model for use with liquid metals by adding a cohesive term to the energy and pressure and by adjusting the soft sphere configurational term with an adjustable parameter Q, since some liquid transition metals have very large electronic contributions to the heat capacity. The final forms for the pressure and energy are

$$P = \frac{NkT}{V}\left[1+\frac{1}{3}nC_n\rho^{n/3}(\varepsilon/kT)+\frac{1}{18}Qn(n+4)\rho^{n/9}(\varepsilon/kT)^{1/3} - m\rho^m(\varepsilon/kT)\right]$$

$$E = NkT\left[\frac{3}{2}+C_n\rho^{n/3}(\varepsilon/kT)+\frac{1}{6}Q(n+4)\rho^{n/9}(\varepsilon/kT)^{1/3} - \rho^m(\varepsilon/kT)\right]+E_{coh}$$

where ε, σ, n, m and Q are treated as adjustable parameters. The fitting procedure involves first a choice of parameters n, m, and Q, and then determination of ε and σ by means of the conditions

$$P(V_m, T_m) = 0, \; E(V_m, T_m) = h_m$$

where the subscript refers to the liquid state at the melting point and h is enthalpy. Isobars of enthalpy, volume, and sound speed as functions of temperature are then computed and compared with experiment. The parameters n, m, and Q are systematically varied until agreement between calculation and experiment is satisfactory. The value of n should be such that $(n+2)6 \cong \gamma_0$, the Grüneisen coefficient at the normal solid density (predicted by harmonic lattice theory with particle potentials of the form $1/r^n$). In general, $m \cong 1.0$ in accord with the van der Waals EOS, and $Q \cong 1$ except where electronic corrections are clearly needed. The primary emphasis in choosing the parameters is given to fitting enthalpy, volume and sound speed isobars. Young used the model to describe Li, Na, Al, K, Ga, Rb, Nb, Mo, Cs, Ta, Hg, Pb and U. Later, fits were made for V, Ir and Pt. Analytic expressions for the specific enthalpy, isothermal bulk modulus B_T, bulk thermal expansion coefficient α and specific heat at constant volume C_v are given below.

Specific enthalpy:

$$h = NkT\,[\frac{5}{2} + \left(\frac{n}{3}+1\right)C_n\rho^{n/3}\left(\frac{\varepsilon}{kT}\right) + \frac{Q}{18}(n+3)(n+4)\rho^{n/9}\left(\frac{\varepsilon}{kT}\right)^{1/3} - (m+1)\rho^m\left(\frac{\varepsilon}{kT}\right)] + E_{coh}$$

Isothermal bulk modulus:

$$B_T = \frac{NkT}{V}[1 + \frac{n}{3}\left(\frac{n}{3}+1\right)C_n\rho^{n/3}\left(\frac{\varepsilon}{kT}\right) + \frac{Q}{162}n(n+4)(n+9)\rho^{n/9}\left(\frac{\varepsilon}{kT}\right)^{1/3} - m(m+1)\rho^m\left(\frac{\varepsilon}{kT}\right)]$$

Bulk thermal expansion coefficient:

$$\beta = \frac{Nk}{VB_T}\left[1+\frac{Q}{27}n(n+4)\rho^{n/9}\left(\frac{\varepsilon}{kT}\right)^{1/3}\right]$$

Specific heat at constant volume:

$$c_v = Nk\left[\frac{3}{2}+\frac{Q}{9}(n+4)\rho^{n/9}\left(\frac{\varepsilon}{kT}\right)^{1/3}\right]$$

Specific heat c_p , Grüneisen coefficient γ_G and sound speed c can then be calculated from

$$c_p = c_v + \beta^2 VTB_T$$
$$\gamma_G = (\gamma - 1)/\gamma T$$

and

$$c = (VB_T\gamma)^{1/2}$$

where $\gamma = c_p/c_v$.

The results for the fits are shown in tables 12.9.1 - 12.9.3. N is Avogadros number divided by the mass per mole. The fits for V, Nb, Ta, Mo, Ir, Pt, Pb and U were made using enthalpy, volume and temperatures obtained from isobaric expansion data. Unfortunately, a systematic error in the temperature measurements from the apparatus was discovered, resulting in specific heats about 14% too large. The fits for Hg, Al, Ga and the alkali metals were unaffected however. Comparison of the predicted critical parameters calculated with the soft sphere model for the affected metals with other measurements suggests that some of the calculated critical temperatures may be too low.

The soft sphere model is a significant improvement over the hard sphere model. The model does poorly in predicting the critical volume and compressibility ratio $Z_c = P_c V_c / T_c$ for alkali metals. A further weakness is the essentially ad hoc nature of the correction for electronic contributions to the heat capacity by means of a multiplicative constant.

Table 12.9.1 Soft sphere parameters and predicted critical point parameters for the alkali elements.

Element	Li	Na	K	Rb	Cs
n	4	8	7	8	7
m	0.7	0.58	0.7	0.7	0.7
Q	1.30	0.95	1.0	1.0	0.90
$N\varepsilon$(J/kg)	1.73095×10^{9}	1.62758×10^{7}	1.34066×10^{7}	4.51764×10^{6}	3.3846×10^{6}
$N\sigma^3/\sqrt{2}(\mathrm{m}^3/\mathrm{kg})$	1.1232×10^{-5}	1.8660×10^{-4}	1.5852×10^{-4}	1.1700×10^{-4}	7.2065×10^{-5}
E_{coh} (J/kg)	2.303×10^{7}	4.659×10^{6}	2.282×10^{6}	9.463×10^{5}	5.723×10^{5}
C_n	12.669	6.3755	6.6658	6.3755	6.6658
P_c (MPa)	114.28	30.367	22.456	15.467	13.635
V_c/V_0	5.8630	5.8486	4.8255	4.9024	5.4017
T_c (K)	3741.2	2428.8	2194.7	1995.3	2018.1
Z_c	0.28041	0.20928	0.27075	0.25628	0.30370
$V_0\ (\mathrm{m}^3/\mathrm{kg})$	1.876×10^{-3}	1.035×10^{-3}	1.166×10^{-3}	6.560×10^{-4}	5.206×10^{-4}

Table 12.9.2 (a) Soft sphere parameters and predicted critical point parameters for transition elements.

Element	V	Nb	Ta	Mo
n	8	6	8	8
m	1.1	1.0	1.1	1.1
Q	2.1	2.1	2.2	2.1
$N\varepsilon$(J/kg)	1.34124×10^{8}	2.47502×10^{8}	5.79905×10^{7}	9.0273×10^{7}
$N\sigma^3/\sqrt{2}(\mathrm{m}^3/\mathrm{kg})$	2.46776×10^{-5}	7.1185×10^{-6}	9.3630×10^{-6}	1.4743×10^{-5}
E_{coh} (J/kg)	1.002×10^{7}	7.764×10^{6}	4.319×10^{6}	6.860×10^{6}
C_n	6.3755	7.2233	6.3755	6.3755
P_c (MPa)	920.24	962.68	998.75	970.20
V_c/V_0	3.7482	4.2343	3.9279	4.4294
T_c (K)	6395.9	9952.2	9284.1	8002.0
Z_c	0.53960	0.53370	0.54885	0.60634
$V_0\ (\mathrm{m}^3/\mathrm{kg})$	1.633×10^{-4}	1.166×10^{-4}	5.968×10^{-5}	9.784×10^{-5}

Table 12.9.2 (b) Soft sphere parameters and predicted critical point parameters for transition elements.

Element	Ir	Pt	Hg
n	12	5	15
m	1.2	1.0	1.5
Q	1.4	1.8	0.65
$N\varepsilon$(J/kg)	1.89318×10^{7}	4.00540×10^{8}	1.86418×10^{6}
$N\sigma^3/\sqrt{2}\left(\mathrm{m^3/kg}\right)$	1.4430×10^{-5}	8.1615×10^{-7}	2.9152×10^{-5}
E_{coh} (J/kg)	3.463×10^{6}	2.895×10^{6}	3.061×10^{5}
C_n	6.0687	8.4422	6.0792
P_c(MPa)	950.13	949.20	178.20
V_c/V_0	4.0037	4.5471	3.4614
T_c (K)	10335	9285.5	2074.4
Z_c	0.37707	0.50849	0.53141
$V_0\left(\mathrm{m^3/kg}\right)$	4.431×10^{-5}	4.661×10^{-5}	7.407×10^{-5}

Table 12.9.3. Soft sphere parameters and predicted critical point parameters for the group 3a elements Al and Ga, the group 4a element Pb and the actinide U.

Element	Al	Ga	Pb	U
n	8	8.8	11	8
m	0.7	0.75	1.05	1.0
Q	1.40	0.95	0.85	2.0
$N\varepsilon$(J/kg)	5.38055×10^{7}	1.61503×10^{7}	4.72852×10^{6}	2.08514×10^{7}
$N\sigma^3/\sqrt{2}\left(\mathrm{m^3/kg}\right)$	7.1220×10^{-5}	3.6057×10^{-5}	2.6820×10^{-5}	9.2172×10^{-6}
E_{coh} (J/kg)	1.220×10^{7}	3.901×10^{6}	9.414×10^{5}	2.197×10^{6}
C_n	6.3755	6.2418	6.0876	6.3755
P_c (MPa)	182.02	254.20	225.95	416.11
V_c/V_0	6.3706	4.6849	3.7037	4.6054
T_c (K)	5726.5	6322.1	5157.5	6617.8
Z_c	0.24352	0.26759	0.35656	0.43444
$V_0\left(\mathrm{m^3/kg}\right)$	3.706×10^{-4}	1.694×10^{-4}	8.818×10^{-5}	5.240×10^{-5}

References

1. *LASL Shock Hugoniot Data,* S. P. Marsh, ed., University of California Press (1980)

2. M. Ross, "Physics of Dense Fluids", in *High Pressure Chemistry and Biochemistry,* R. Van Eldik and J. Jonas, eds., pp 9-49, Reidel (1987)

3. F. E. Walker, F. G. Walker, and J. B. Walker, "Calculation of Hugoniot Values from Atomic Properties", J. Appl. Phys. **60**(1), 3876 - 3881 (1986)

4. U. Walzer, "Calculations of the Hugoniot Pressure and Pressure Derivative of the Bulk Modulus for Transition Metals", High Temp. - High Press. **19**, 161-176 (1987)

5. R. W. Richtmeyer and K. W. Morton, *Difference Methods for Initial-Value Problems,* (2nd Ed), pg 308, Wiley, (1967)

6. R. G. McQueen & S. P. Marsh, "Equation of State for Nineteen Elements from Shock-Wave Measurements to Two Megabars", J. Appl. Phys. **31**, 1253-1269 (1960)

7. M. H. Rice, R. G. McQueen, J. M. Walsh, "Compression of Solids by Strong Shock Waves", in *Solid State Physics* **6**, 1-63 (F. Seitz and D. Turnbull, eds.) Academic Press (1958)

8. S. A. Novikov and A. V. Chernov, "Determination of the Spall Strength from Measured Values of the Specimen Free-Surface Velocity", J. Appl. Mech. Tech. Phys. **23**, 703 - 705 (USSR) (1982)

9. *High-Velocity Impact Phenomena,* R. Kinslow, ed., Academic Press (1970)

10. P. W. Bridgman, *A condensed Collection of Thermodynamic Formulas,* Dover (1961)

11. J. C. Slater, *Introduction to Chemical Physics,* McGraw-Hill (1939)

12. E. T. Whittaker, *Analytical Dynamics,* 4th Ed., (Chap VII, pp 181-182), Cambridge University Press (1964)

13. J. H. Tillotson, *Metallic Equations of State for Hypervelocity Impact*, General Atomic Report No. GA-3216 (1962)

14. L. A. Glenn, D. A. Young, "Dynamic Loading of the Structural Wall in a Lithium-Fall Fusion Reactor", Nucl. Eng. Design **54**, 1-16 (1979)

15. D. A. Young, *A soft Sphere Model for Liquid Metals*, Lawrence Livermore National Laboratory, University of California Report No. UCRL-52352 (1977)

16. W. G. Hoover, G. Stell, E. Goldmark and G. D. Degani, "Generalized van der Waals Equation of State", J. Chem. Phys. **63**, 5434-5438 (1975)

Additional Resources

Journal Articles:

1. S. B. Segletes, "Thermodynamic Stability of the Mie-Gruneisen Equation of State, and its Relevance to Hydrodynamic Computations", J. Appl. Phys. **70**(5), 2489-2499 (1991)

2. R. G. McQueen, J. W. Hopson and J. N. Fritz, "Optical Technique for Determining Rarefaction Wave Velocities at Very High Pressures", Rev. Sci. Instrum. **53**(2), 245-250 (1982)

3. V. F. Anisichkin, "Generalized Shock Adiabats and Zero Isotherms of Elements", Comb. Explo. and Shock Waves **15**(2), 245-250 (USSR) (1979)

4. R. A. Graham and J. R. Asay, "Measurement of Wave Profiles in Shock-Loaded Solids", High Temp.-High Press. **10**, 355-390 (1978)

5. R. A. Graham and J. R. Asay, "Measurement of Wave Profiles in Shock-Loaded Solids", High Temp.-High Press. **10**, 355-390 (1978)

6. L. V. Al'tshuler, N. N. Kalitkin, L. V. Kuz'mina and B. S. Chekin, "Shock Adiabats For Ultrahigh Pressures", Sov. Phys. JETP **45**(1), 167-171 (1977)

7. G. E. Duvall and R. A. Graham, "Phase Transitions Under Shock-Wave Loading", Rev. Mod. Phys. **49**(3), 523-579 (1977)

8. L. V. Al'tshuler, "Use of Shock Waves in High-Pressure Physics", Sov. Phys. Uspekhi **8**(1), 52-91 (1965)

9. G. E. Duvall, "Concepts of Shock Wave Propagation", Bull. Seis. Soc. Am. **52**(4), 869-893 (1962)

10. G. E. Duvall, "Shock Waves in the Study of Solids", Appl. Mech. Rev. **15**(11), 849-854 (1962)

11. L. V. Al'tshuler, S. B. Kormer, M. I. Brazhnik, L. A. Vladimirov, M. P. Speranskaya and A. I. Funtikov, "The Isentropic Compressibility of Aluminum, Copper, Lead and Iron at High Pressures", Sov. Phys. JETP **11**(4), 766-213 (1960)

12. J. M. Walsh, M. H. Rice, R. G. McQueen and F. L. Yarger, "Shock-Wave Compression of Twenty-Seven Metals. Equation of State", Phys. Rev. **108**(2), 196-216 (1957)

13. J. S. Dugdale and D. K. C. MacDonald, "The Thermal Expansion of Solids", Phys. Rev. 89(4), 832-834 (1953)

Books:

14. R. A. Graham, *Solids Under High-Pressure Shock compression,* Springer-Verlag (1993)

15. *High Pressure Shock Compression,* J. R. Asay & M. Shahinpoor (eds.), Springer-Verlag (1992)

16. S. Eliezer, A. Ghatak and H. Hora, *An Introduction to Equations of State,* Cambridge University Press (1986)

17. M. A. Liberman and A. L. Velikovich, *Physics of Shock Waves in Gases and Plasmas,* Springer-Verlag (1985)

18. J. G. Kirkwood, *Shock and Detonation Waves,* Gordon and Breach (1967)

19. Ye. V. Stupochenko, S. A. Losev and A. I. Osipov, *Relaxation in Shock Waves*, Springer-Verlag (1967)

20. Ya. B. Zel'dovich and Yu. P. Raizer, *Physics of Shock-Waves and High Temperature Hydrodynamic Phenomena*, Academic Press (1966)

21. J. N. Bradley, *Shock Waves in Chemistry and Physics*, Methuen (1962)

22. R. Courant and K. D. Friedrichs, *Supersonic Flow and Shock Waves*, Interscience (1948)

Conference Proceedings:

23. *Shock Waves in Condensed Matter-1981*, W. J. Nellis, L. Seaman, R. A. Graham, eds., American Institute of Physics (1982)

24. *Shock Waves in Condensed Matter-1983*, J. R. Asay, R. A. Graham, G. K. Straub, eds., North-Holland (1984)

25. *Shock Waves in Condensed Matter*, Y. M. Gupta, ed., Plenum (1986)

26. *Shock Waves in Condensed Matter 1987*, S. C. Shmidt, N. C. Holmes, eds., North-Holland (1988)

27. *Shock Waves in Condensed Matter-1989*, S. C. Shmidt, J. N. Johnson, L. W. Davison, eds., North-Holland (1990)

28. *Shock Waves in Condensed Matter-1991*, S. C. Schmidt, R. D. Dick, J. W. Forbes, D. G. Tasker, eds., North-Holland (1992)

29. *Hypervelocity Impact*, Int. J. Impact Eng. 5 (1-4), 1987

30. *Hypervelocity Impact*, Int. J. Impact Eng. 10 (1-4), 1990

31. *Physics of High Energy Density*, P. Caldirola, ed., Academic Press (1971)

The book by Asay and Shahinpoor contains an extensive bibliography with 252 listings, and an appendix with up-to-date Hugoniot fits and Grüneisen coefficients for many materials from GMX-6 at LANL. It is part of a series titled "High-Pressure Shock Compression of Condensed Matter", and is recommended for those wishing a very broad view of the field in its present state. The material covers basic principles, experimental and diagnostic technique, equation of state, inelastic constitutive relations, dynamic fracture, large deformation wave codes, micromechanical considerations in shock compression of solids, and the influence of shock-wave deformation on the structural/property behavior of materials. The level of mathematical treatment varies greatly, depending on the section. In some chapters the treatment is rather condensed and assumes a great deal of background. In others, there is a great deal of detail, but the treatment is rather advanced (eg; much tensor analysis.)

The book by Courant and Friedrichs is quite rigorous, and suitable for those with a strong math background. It was long considered the best source. For many, however, the physics gets lost in the rigor. The work by Zel'dovich and Raizer is a two volume set, edited from the Russian by W. D. Hayes and R. F. Probstein. It has often been used as a text. The scope is enormous, and the treatment rather condensed as a result. Many beginning readers find it difficult to use. The book by Bradley deals mainly with physical phenomena in shock tubes. It focuses a good deal on chemistry. Item 18 is the collected papers of J. G. Kirkwood.

The book by Eliezer, et al. is rather directed toward theoreticians. The book by Stupochenko, et al is concerned mainly with dissipative mechanisms for shock waves in gases. The contents of the book by Liberman and Velikovich are well reflected by its title. Graham's book is relatively modest in the mathematical burden and focused primarily on processes occurring behind shock fronts in solids. It should be mentioned that the book in the reference list, edited by R. Kinslow is an earlier collection similar to that of Asay and Shahinpoor.

The first six conference proceedings listed are those from the on-going series sponsored by the American Physical Society. The next two contain reports of work largely supported by funds from the Strategic Defense Initiative. The last is the proceedings of the Enrico Fermi International School of Physics, dealing with high pressure physics work using shock waves.

The Journal articles are only a sample of what is available. The article by Segletes deals with practical difficulties seen with the Mie-Grüneisen EOS when using computer simulation codes. The contents of the remainder are generally described by their titles.

Appendix A: Solution of the Cubic with Real Coefficients

Assume the cubic with real coefficients in the standard form

$$x^3 + ax^2 + bx + c = 0$$

Calculate

$$h = -\frac{c}{2} + \frac{ab}{6} - \frac{a^3}{27}$$

$$w_1^2 = -3\left(a^2b^2 - 4a^3c + 18abc - 27c^2\right) + 12b^3$$

$$w_2^3 = h + \frac{w_1}{18} \quad , \; w_3 = \frac{a^2/9 - b/3}{w_2} \, , \; x = -\frac{a}{3} + w_2 + w_3$$

Note that if $w_2 = 0$, the method fails. In general, x, w_2, w_3 are complex numbers. We write $w_2^3 = re^{i\theta}$ then from complex variable analysis we have

$$w_2^{(k)} = r^{\frac{1}{3}}\left[\cos\left(\frac{\theta + 2\pi k}{3}\right) + i\,\sin\left(\frac{\theta + 2\pi k}{3}\right)\right], \quad k = 0, 1, 2$$

and

$$w_3^{(k)} = \frac{a^2/9 - b/3}{w_2^{(k)}}, \quad x^{(k)} = -\frac{a}{3} + w_2^{(k)} + w_3^{(k)}$$

so that

$$x^{(k)} = -\frac{a}{3} + w_2^{(k)} + \frac{a^2/9 - b/3}{w_2^{(k)}} \bullet \frac{w_2^{(k)*}}{w_2^{(k)*}}$$

where the superscript * denotes complex conjugate.

$$x^{(k)} = -\frac{a}{3} + w_2^{(k)} + \left(\frac{a^2}{9} - \frac{b}{3}\right)\frac{w_2^{(k)*}}{|w_2^{(k)}|^2}$$

but $\left|w_2^{(k)}\right|^2 = r^{\frac{2}{3}}$ so that

$$\frac{w_2^{(k)*}}{|w_2^{(k)}|^2} = \frac{1}{r^{\frac{1}{3}}}\left[\cos\left(\frac{\theta+2\pi k}{3}\right) - i\sin\left(\frac{\theta+2\pi k}{3}\right)\right]$$

and

$$x^{(k)} = -\frac{a}{3} + r^{\frac{1}{3}}\left[\cos\left(\frac{\theta+2\pi k}{3}\right) + i\sin\left(\frac{\theta+2\pi k}{3}\right)\right]$$
$$+\left(\frac{a^2}{9} - \frac{b}{3}\right)\frac{1}{r^{\frac{1}{3}}}\left[\cos\left(\frac{\theta+2\pi k}{3}\right) - i\sin\left(\frac{\theta+2\pi k}{3}\right)\right]$$

or

$$x^{(k)} = \left[-\frac{a}{3} + \left\{r^{\frac{1}{3}} + \left(\frac{a^2}{9} - \frac{b}{3}\right)\frac{1}{r^{\frac{1}{3}}}\right\}\cos\left(\frac{\theta+2\pi k}{3}\right)\right]$$
$$+i\left[r^{\frac{1}{3}} - \left(\frac{a^2}{9} - \frac{b}{3}\right)\frac{1}{r^{\frac{1}{3}}}\right]\sin\left(\frac{\theta+2\pi k}{3}\right)$$

where $k = 0,\ 1\ 2$ and $r = |w_2^3|$

Consider a couple of illustrative examples

Example 1.

$$x^3 + x^2 + x + 1 = 0 \qquad (a,\ b,\ c = 1)$$

$$h = -\frac{1}{2} + \frac{1}{6} - \frac{1}{27} = -\frac{10}{27}$$
$$w_1^2 = -3(1-4+18-27)+12 = 48\,, \quad w_1 = 4\sqrt{3}$$
$$w_2^3 = -\frac{10}{27} + \frac{4\sqrt{3}}{18} = \frac{6\sqrt{3}-10}{27} = re^{i\theta}$$

$$r = \frac{6\sqrt{3}-10}{27}\,,\ \theta = 0\,,\ r^{\frac{1}{3}} = \frac{\left(6\sqrt{3}-10\right)^{\frac{1}{3}}}{3} = 0.2440169359$$
$$w_2^{(k)} = r^{\frac{1}{3}}\left[\cos\left(\frac{2\pi k}{3}\right) + i\sin\left(\frac{2\pi k}{3}\right)\right]$$

$$w_2^{(0)} = r^{\frac{1}{3}},\ w_2^{(1)} = r^{\frac{1}{3}}\left(\frac{-1+i\sqrt{3}}{2}.\right),\ w_2^{(2)} = r^{\frac{1}{3}}\left(\frac{-1-i\sqrt{3}}{2}\right)$$

$$w_3^{(0)} = -\frac{2}{9r^{\frac{1}{3}}},\ w_3^{(1)} = \frac{\left(1+i\sqrt{3}\right)}{9r^{\frac{1}{3}}},\ w_3^{(2)} = \frac{\left(1-i\sqrt{3}\right)}{9r^{\frac{1}{3}}}$$

$$\left(\frac{a^2}{9}-\frac{b}{3}\right) = -\frac{2}{9}$$

$$x^{(k)} = \left[-\frac{1}{3}+\left\{r^{\frac{1}{3}}-\frac{2}{9r^{\frac{1}{3}}}\right\}\cos\left(\frac{2\pi k}{3}\right)\right]+i\left[r^{\frac{1}{3}}+\frac{2}{9r^{\frac{1}{3}}}\right]\sin\left(\frac{2\pi k}{3}\right)$$

and

$$\left\{r^{\frac{1}{3}}-\frac{2}{9r^{\frac{1}{3}}}\right\} = -\frac{2}{3},\ \left[r^{\frac{1}{3}}+\frac{2}{9r^{\frac{1}{3}}}\right] = \frac{2}{\sqrt{3}}$$

so that

$$x^{(k)} = \left[-\frac{1}{3}-\frac{2}{3}\cos\left(\frac{2\pi k}{3}\right)\right]+i\frac{2}{\sqrt{3}}\sin\left(\frac{2\pi k}{3}\right)$$

and since

$$\cos\left(\frac{2\pi}{3}\right) = -\cos\left(\frac{\pi}{3}\right) = -\frac{1}{2},\ \sin\left(\frac{2\pi}{3}\right) = \sin\left(\frac{\pi}{3}\right) = \frac{\sqrt{3}}{2}$$

$$\cos\left(\frac{4\pi}{3}\right) = -\cos\left(\frac{\pi}{3}\right) = -\frac{1}{2},\ \sin\left(\frac{4\pi}{3}\right) = -\sin\left(\frac{\pi}{3}\right) = -\frac{\sqrt{3}}{2}$$

we have

$$x^{(0)} = -1$$

$$x^{(1)} = \left[-\frac{1}{3}-\frac{2}{3}\left(-\frac{1}{2}\right)\right]+i\left[\frac{2}{\sqrt{3}}\left(\frac{\sqrt{3}}{2}\right)\right] = i$$

$$x^{(2)} = \left[-\frac{1}{3}-\frac{2}{3}\left(-\frac{1}{2}\right)\right]+i\left[\frac{2}{\sqrt{3}}\left(-\frac{\sqrt{3}}{2}\right)\right] = -i$$

Example 2:

$x^3 - x^2 - 2x + 1 = 0 \qquad (a = -1, b = -2, c = 1)$

$$h = -\frac{1}{2} + \frac{2}{6} + \frac{1}{27} = -\frac{7}{54}, \quad w_1^2 = -147, \quad w_2^3 = \frac{-7 + 3i\sqrt{147}}{54},$$

$$r = \frac{\sqrt{343}}{27}, \quad r^{\frac{1}{3}} = \frac{\sqrt{7}}{3}, \quad \tan\theta = -\frac{3\sqrt{147}}{7} = -5.196152423$$

$$\theta = -79.10660535^\circ \quad \{Re(w_2^3) < 0, \text{ hence add } 180^\circ:$$
$$\theta = 100.8933946^\circ = 1.760921929 \text{ Rad}\}$$

$$\left(\frac{a^2}{9} - \frac{b}{3}\right) = \frac{7}{9}, \left(\frac{a^2}{9} - \frac{b}{3}\right)\frac{1}{r^{\frac{1}{3}}} = \frac{7}{9} \bullet \frac{3}{\sqrt{7}} = \frac{\sqrt{7}}{3}$$

We thus have $r^{\frac{1}{3}} - \left(\frac{a^2}{9} - \frac{b}{3}\right)\frac{1}{r^{\frac{1}{3}}} = 0$ and the roots are all real.

Now

$$r^{\frac{1}{3}} + \left(\frac{a^2}{9} - \frac{b}{3}\right)\frac{1}{r^{\frac{1}{3}}} = \frac{2}{3}\sqrt{7}$$

We thus have

$$x^{(k)} = \left[\frac{1}{3} + \frac{2}{3}\sqrt{7}\cos\left(\frac{\theta + 2\pi k}{3}\right)\right]$$

and

$$\cos\left(\frac{\theta}{3}\right) = 0.8326204337$$
$$\cos\left(\frac{\theta + 2\pi}{3}\right) = -0.8959532195$$
$$\cos\left(\frac{\theta + 4\pi}{3}\right) = 0.0633327857$$

We thus obtain the roots:

$$x^{(0)} = 1.801937736$$
$$x^{(1)} = -1.246979603$$
$$x^{(2)} = 0.4450418672$$

Code for Solving the Cubic With Real Coefficients

The following code implements the development above, and solves the arbitrary cubic with real coefficients.

```
100 REM General Cubic Equation Solver (real coefficients)
110 REM Form: X**3+A*X**2+B*X+C = 0
150 PRINT "This code analytically solves in the"
160 PRINT "complex plane for the roots of the"
170 PRINT "arbitrary cubic equation with real"
180 PRINT "coefficients":PRINT
190 PRINT "Results < 3.E-7 are treated as zero"
200 PRINT
210 PRINT "The assumed form is X**3+A*X**2+B*X+C = 0"
230 PRINT
240 INPUT "A= ";A
250 INPUT "B= ";B
260 INPUT "C= ";C
270 PI = 3.1415926
280 H = A*B/6-A*A*A/27-C/2
290 W1 = 12*A*A*A*C-3*A*A*B*B-54*A*B*C
300 W1 = W1+81*C*C+12*B*B*B
310 IF W1 <= 0 GOTO 350
320 W2 = SQR(W1)
330 RP = W2/18+H
340 IP = 0: GOTO 380
350 W2 = SQR(-W1)
360 IP = W2/18
370 RP = H
380 R4 = A*A/9-B/3
390 R5 = SQR(RP*RP+IP*IP)
400 TH = ATN(IP/RP)
410 IF RP<0 THEN TH = TH+PI
420 R6 = R5^(1/3)
430 C1 = R6+R4/R6
440 C2 = R6-R4/R6
450 X0 = -A/3+C1*COS(TH/3)
460 Y0= C2*SIN(TH/3)
470 X1 = -A/3+C1*COS((TH+2*PI)/3)
480 Y1 = C2*SIN((TH+2*PI)/3)
```

```
490 X2 = -A/3+C1*COS((TH+4*PI)/3)
500 Y2 = C2*SIN((TH+4*PI)/3)
510 IF ABS(X0) < .0000003 THEN X0 = 0
520 IF ABS(Y0) < .0000003 THEN Y0 = 0
530 IF ABS(X1) < .0000003 THEN X1 = 0
540 IF ABS(Y1) < .0000003 THEN Y1 = 0
540 IF ABS(X2) < .0000003 THEN X2 = 0
540 IF ABS(Y2) < .0000003 THEN Y2 = 0
570 PRINT:PRINT "Root";TAB(10);"Re(x)";TAB(28);"Im(x)":PRINT
600 PRINT "1: ";TAB(8);X0;TAB(24);Y0
610 PRINT "2: ";TAB(8);X1;TAB(24);Y1
620 PRINT "3: ";TAB(8);X2;TAB(24);Y2
630 END
```

Application to inversion of the ratio of polynomials EOS

With the coefficients used in ratio of polynomials model for the hydrogen equation of state, it was found that the k=0 solution is the one of interest, accordingly we can simplify to

$$x^{(0)} = -\frac{a}{3}+\left\{r^{\frac{1}{3}}+\left(\frac{a^2}{9}-\frac{b}{3}\right)\frac{1}{r^{\frac{1}{3}}}\right\}\cos\left(\frac{\theta}{3}\right)$$
$$+i\left[r^{\frac{1}{3}}-\left(\frac{a^2}{9}-\frac{b}{3}\right)\frac{1}{r^{\frac{1}{3}}}\right]\sin\left(\frac{\theta}{3}\right)$$

where $r = |w_2^{\frac{3}{2}}|$, and

$$\theta = \tan^{-1}\left(\frac{\mathrm{Im}(w_2^{\frac{3}{2}})}{\mathrm{Re}(w_2^{\frac{3}{2}})}\right)$$

Only real roots are of interest. We can thus drop the second term and further simplifiy to

$$x^{(0)} = -\frac{a}{3} + \left\{ r^{\frac{1}{3}} + \left(\frac{a^2}{9} - \frac{b}{3} \right) \frac{1}{r^{\frac{1}{3}}} \right\} \cos\left(\frac{\theta}{3} \right)$$

where $r = |w_2^3|$, and

$$\theta = \tan^{-1}\left(\frac{\mathrm{Im}(w_2^3)}{\mathrm{Re}(w_2^3)} \right)$$

For $\mathrm{Re}(w_2^3) < 0$, add π to θ

A code listing in FORTRAN follows implementing this calculation.

```
SUBROUTINE INVCUBE(A, B, C,X0)
H = A*B/6.-A*A*A/27.-C/2
W12 = 12.*A*A*A*C-3.*A*A*B*B-54.*A*B*C
W12 = W12+81.*C*C+12.*B*B*B
IF(W12.LE.0) GO TO 10
W1 = SQRT(W12)
RPW23 = W1/18.+H
FIPW23 = 0.
GO TO 20
10 W1 = SQRT(-W12)
FIPW23 = W1/18.
RPW23 = H
20 R14 = A*A/9.-B/3.
R = SQRT(RPW23*RPW23+FIPW23*FIPW23)
THETA = ATANF(FIPW23/RPW23)
IF(RPW23.LT.0) THEN
THETA = THETA+3.1415926535898
ENDIF
R13P = R**(1./3.)
RPW20 = R13P*COSF(THETA/3.)
DENM = R13P*R13P
RPW30=R14*RPW20/DENM
X0 = RPW20+RPW30-A/3.
RETURN
END
```

Appendix B: Multi-branched EOS for Lithium

Lines 110 - 250 are a driver for demonstration purposes. The remaining lines are the coding of the EOS

```
110 REM Units: Rho (g/cm3), c(km/sec), P(GPa), E (MJ/kg), T(kK),
115 REM R(kJ/kgK)
120 REM Constants are for Lithium
130 G0 = 0.9: RH0 = 0.518: PC = 0.1
140 EC = 23.03: TEX = 0.2: R = 1.198
150 T0 = 0.454: E1 = 14: AEX = 1.4
160 BEX = 1.481: EX1 = 1.5: C = 4.5
170 S = 1.133: TTH = 2/3
180 INPUT "E(MJ/kg) ?", E
190 INPUT "RH(g/cm3) ?", RH
200 GOSUB 1600
210 GOSUB 1000
220 PRINT "P(GPa) =", P
230 PRINT "CS(km/s) =", P
240 PRINT "T(kK) =", T
250 END

1000 IF E>0 THEN 1020
1005 REM error trap
1010 P = 0: CS = 0: T = 0: RETURN
1020 IF RH>=RH0 THEN 1480
1030 IF E>EC THEN 1400
1040 T = (RH/RH0)^TEX*(E/(3*R)+T0)
1050 IF E>E1 THEN 1200
1055 REM Region 3 below E1
1060 P51 = PC*EXP(AEX)*EXP(-BEX*EC/E)
1070 C51 = P51*SQR(BEX*EC)/(RH*E)
1080 P52 = G0*RH0*E*(RH/RH0)^EX1
1090 C52 = SQR(P52*EX1/RH+P52^2/(RH^2*E))
1100 IF P51>P52 THEN 1120
1110 P5 = P51: C5 = C51: GOTO 1130
1120 P5 = P52: C5 = C52
1130 P4 = RH0*C^2*(RH/RH0-1)+G0*RH0*E
1140 IF P4<=0 THEN 1160
1150 C4 = SQR(C^2+P4*G0*RH0/RH^2): GOTO 1170
```

```
1160 C4 = C
1170 IF P4>P5 THEN 1190
1180 P = P5: CS = C5: RETURN
1190 P = P4: CS = C4: RETURN
1195 REM Interpolation zone of region 3
1200 P41 = RH0*C^2*(RH/RH0-1)+G0*RH0*E1
1210 DPR41 = C^2: DPE41 = 0
1220 P511 = PC*EXP(AEX)*EXP(-BEX*EC/E)
1230 DPR511 = 0: DPE511 = P511*BEX*EC/E^2
1240 P512 = G0*RH0*E*(RH/RH0)^EX1
1250 DPR512 = P512*EX1/RH: DPE512 = P512/E
1260 IF P511>P512 THEN 1280
1270 P51 = P511: DPR51 = DPR511: DPE51 = DPE511: GOTO 1290
1280 P51 = P512: DPR51 = DPR512: DPE51 = DPE512
1290 P2 = G0*RH0*EC*(RH/RH0)^EX1
1300 DPR2 = P2*EX1/RH: DPE2 = 0
1310 IF P41>P51 THEN 1360
1320 P = P51+(E-E1)*(P2-P51)/(EC-E1)
1330 DPR = DPR51+(E-E1)*(DPR2-DPR51)/(EC-E1)
1340 DPE = DPE51+(P2-P51)/(EC-E1)+(E-E1)*(DPE2-DPE51)/(EC-E1)
1350 CS = SQR(DPR+P*DPE/RH^2): RETURN
1360 P = P41+(E-E1)*(P2-P41)/(EC-E1)
1370 DPR = DPR41+(E-E1)*(DPR2-DPR41)/(EC-E1)
1380 DPE = DPE41+(P2-P41)/(EC-E1)+(E-E1)*(DPE2-DPE41)/(EC-E1)
1390 CS = SQR(DPR+P*DPE/RH^2): RETURN
1395 REM Region 2 (gas)
1400 P = TTH*(E-EC)*RH+(G0*RH0*E-TTH*
        (E-EC)*RH0)*(RH/RH0)^EX1
1410 DPR = TTH*(E-EC)*FEX+(G0*RH0*E-TTH*
        (E-EC)*RH0)*FEX)*EX1*(RH/RH0)^EX1/RH
1420 DPE = TTH*RH*FEX+TTH*(E-EC)*RH*FPEX
1430 DPE = DPE+(G0*RH0-TTH*RH0*FEX
        -TTH*(E-EC)*RH0*FPEX)*(RH/RH0)^EX1
1440 CS = SQR(DPR+P*DPE/RH^2)
1450 TG = (1/2+7/2*EC/E)*(E-EC)/(6*R)
1460 TS = E/(3*R)+T0
1470 T = TG+(RH/RH0)^TEX*(TS-TG): RETURN
1475 REM Region 1 (Grüneisen model)
1480 X = 1-RH0/RH
1490 PH = RH0*C^2*X/(1-S*X)^2
1500 P = PH*(1-G0*X/2)+G0*RH0*E
```

```
1510 DPX = RH0*C^2*(1-G0*X+2*X*(1-G0*X/2)/(1-S*X))/(1-S*X)^2
1520 DPR = DPX*RH0/RH^2
1530 DPE = G0*RH0
1540 CS = SQR(DPR+P*DPE/RH^2)
1550 E00 = -3*R*T0
1560 E0 = E00+(G0*E00)*X+.5*(C^2+G0^2*E00)*X^2
1570 E0 = E0+(4*S*C^2+G0^3*E00)*X^3/6
1580 T = (E-E0)/(3*R)
1590 RETURN
1600 Y = LOG(E/EC)
1610 FEX = 1-.7*EXP(-2*(Y-1.233)^2)
1620 FEX = FEX-.7*EXP(-.5*(Y-4.468)^2)
1630 FPEX = 2.8*(Y-1.233)*EXP(-2*(Y-1.233)^2)
1640 FPEX = FPEX+.7*(Y-4.468)*EXP(-.5*(Y-4.468)^2)
1650 FPEX = FPEX/E
1660 RETURN
```

Appendix C: Useful Computer Routines for Plane Shock Waves

This appendix contains a number of useful approximate calculations using the approximation of the reshock Hugoniot by the reflection of the principal Hugoniot and the approximation of a release isentrope by either the principal Hugoniot for the case of an overtaking rarefaction or the reflection of the principal Hugoniot in the case of a rear-facing rarefaction caused by arrival of the shock at a material with lower shock impedance. To simplify nomenclature, $H(u)$ is the pressure on the Hugoniot (ie; $H(u)$ refers to the principal Hugoniot, $H(u_I-u)$ refers to the principal Hugoniot reflected in the origin and translated by u_I.) The subscript on the Hugoniot identifies the material it belongs to.

C.1 Shocks Through a Stack of Materials

Figure C.1 illustrates the analysis. Continuity of pressure across an interface gives

$$
\begin{array}{lr}
H_1(u_I-u) = H_2(u_1) & \text{(point 1)} \\
H_2(2u_1-u_2) = H_3(u_2) & \text{(point 2)} \\
H_3(2u_2-u_3) = H_4(u_3) & \text{(point 3)} \\
H_4(2u_3-u_4) = H_5(u_4) & \text{(point 4)}
\end{array}
$$

The figure is drawn for the case $Z_4 > Z_5 > Z_2 > Z_1 > Z_3$ (where Z is shock impedance) but the calculation is not restricted to that case. We assume a linear D,u relation is adequate to describe each Hugoniot. We have

$$
\begin{aligned}
\rho_{01}\left[c_{01}+s_1(u_I-u_1)\right](u_I-u_1) &= \rho_{02}(c_{02}+s_2u_1)u_1 \\
\rho_{02}\left[c_{02}+s_2(2u_1-u_2)\right](2u_1-u_2) &= \rho_{03}(c_{03}+s_3u_2)u_2 \\
\rho_{03}\left[c_{03}+s_3(2u_2-u_3)\right](2u_2-u_3) &= \rho_{04}(c_{04}+s_4u_3)u_3 \\
\rho_{04}\left[c_{04}+s_4(2u_3-u_4)\right](2u_3-u_4) &= \rho_{05}(c_{05}+s_5u_4)u_4
\end{aligned}
$$

The above set of equations can be solved successively for u_1, u_2, u_3, and u_4 First simplify nomenclature a bit:

$$a_i \equiv \rho_{0i} c_{0i}\ , \quad b_i \equiv \rho_{0i} s_i \qquad i = 1, 2, 3, 4$$

We have

$$a_1 (u_I - u_1) + b_1 (u_I - u_1)^2 = a_2 u_1 + b_2 u_1^2 \tag{C.1}$$
$$a_2 (2u_1 - u_2) + b_2 (2u_1 - u_2)^2 = a_3 u_2 + b_3 u_2^2 \tag{C.2}$$
$$a_3 (2u_2 - u_3) + b_3 (2u_2 - u_3)^2 = a_4 u_3 + b_4 u_3^2 \tag{C.3}$$
$$a_4 (2u_3 - u_4) + b_2 (2u_3 - u_4)^2 = a_5 u_4 + b_5 u_4^2 \tag{C.4}$$

First solve (C.1) for u1:

$$u_1 = \frac{(a_1 + 2b_1 u_I + a_2) - \sqrt{(a_1 + 2b_1 u_I + a_2)^2 - 4(b_1 - b_2)(a_1 + b_1 u_I) u_I}}{2(b_1 - b_2)}$$

For symmetric impact, where $b_1 = b_2$ the solution fails. Symmetry gives $u_1 = u_I/2$ for that case.

Now solve (C.2) for u_2 :

$$u_2 = \frac{(a_2 + 4b_2 u_1 + a_3) - \sqrt{(a_2 + 4b_2 u_1 + a_3)^2 - 8(b_2 - b_3)(a_2 + 2b_2 u_1) u_1}}{2(b_2 - b_3)}$$

If material 2 is identical to material 3 the solution fails since $b_2 = b_3$. Continuity however, then gives $u_2 = u_1$.

Solve (C.3) for u_3 :

$$u_3 = \frac{(a_3 + 4b_3 u_2 + a_4) - \sqrt{(a_3 + 4b_3 u_2 + a_4)^2 - 8(b_3 - b_4)(a_3 + 2b_3 u_2) u_2}}{2(b_3 - b_4)}$$

If material 4 is the same as material 3 the solution fails since $b_3 = b_4$. Continuity gives $u_3 = u_2$.

Solve (C.4) for u_4 :

$$u_4 = \frac{(a_4 + 4b_4 u_3 + a_5) - \sqrt{(a_4 + 4b_4 u_3 + a_5)^2 - 8(b_4 - b_5)(a_4 + 2b_4 u_3)u_3}}{2(b_4 - b_5)}$$

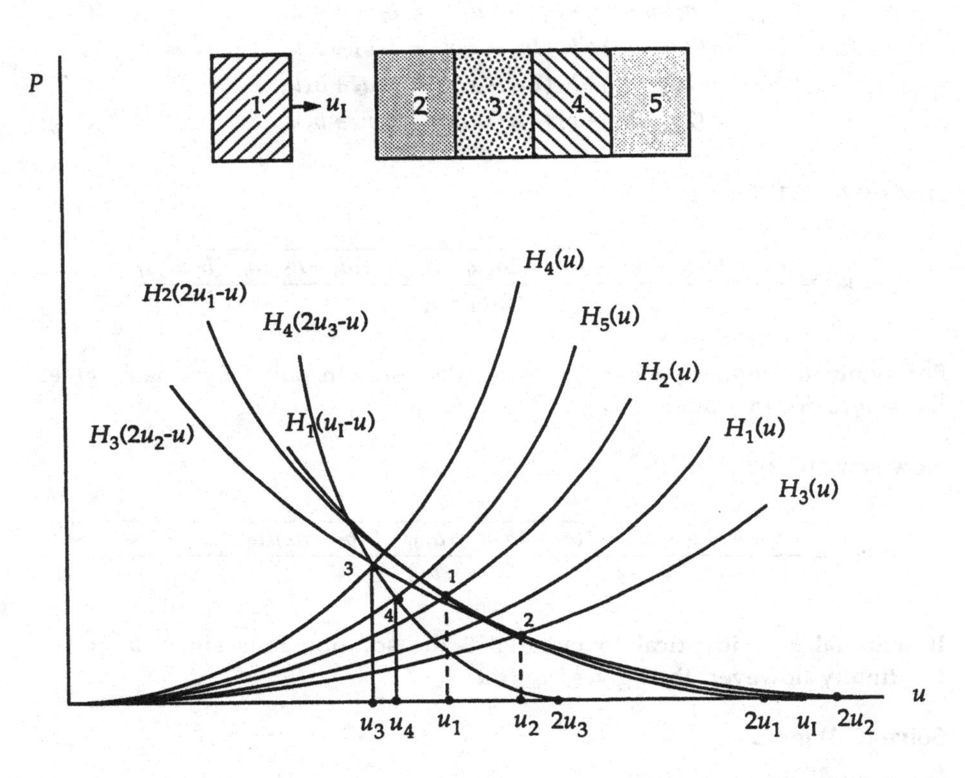

Fig. C.1. Analysis of shock impact of a flyer on a plane stack of materials of infinite width and with flyer thickness sufficient that overtaking rarefactions do not catch up to the shock wave in the stack.

If material 5 is the same as material 4 the solution fails, since $b_4 = b_5$. Continuity then gives $u_4 = u_3$.

It may be seen that the pattern is

$$u_i = \frac{(a_i + 4b_i u_{i-1} + a_{i+1}) - \sqrt{(a_i + 4b_i u_{i-1} + a_{i+1})^2 - 8(b_i - b_{i+1})(a_i + 2b_i u_{i-1})u_{i-1}}}{2(b_i - b_{i+1})}$$

with $u_0 = u_I/2$ for the case $i = 1$. For material i the same as material $i+1$, continuity gives $u_i = u_{i-1}$. (C.5)

An additional special case occurs. suppose two materials in contact have Hugoniot stiffness constants such that the ratio of the constants is the inverse of the ratio of their densities, but whose normal sound speeds do not have the same ratio:

$$\frac{s_{i+1}}{s_i} = \frac{\rho_{0i}}{\rho_{0i+1}} \neq \frac{c_{0i+1}}{c_{0i}}$$

This gives $b_{i+1} = b_i$, $a_{i+1} \neq a_i$ The equation for u_i is then

$$2(a_i + 2b_i u_{i-1})u_{i-1} - (a_i + 4b_i u_{i-1} + a_{i+1})u_i + (b_i - b_{i+1})u_i^2 = 0$$

but for $b_{i+1} = b_i$ the quadratic term drops out and the solution becomes

$$u_i = \frac{2(a_i + 2b_i u_{i-1})u_{i-1}}{(a_i + 4b_i u_{i-1} + a_{i+1})} \qquad (C.6)$$

Now consider the case where the materials are in fact identical, so that $a_i = a_{i+1}$ as well:

$$u_i = \frac{2(a_i + 2b_i u_{i-1})u_{i-1}}{2(a_i + 2b_i u_{i-1})} = u_{i-1}$$

Note that $P_i = a_{i+1}u_i + b_{i+1}u_i^2$.

A code implementing the calculation follows.

```
10 REM Shock Prediction Code
20 REM This code calculates the shock pressure and particle
```

```
30 REM velocities in each material for the problem.
40 REM Material 1 is the impactor for the problem.
50 REM The results assume no overtaking or side
60 REM rarefactions occur.  If the number of materials
70 REM exceeds 10 a dimension statement must be added,
80 REM otherwise an arbitrarily large number of materials
90 REM may be used.  Accuracy degrades progressively
100 REM through the stack, however
110 REM
120 INPUT "How many materials?", NM
130 PRINT "Give the following properties:":PRINT
140 FOR I = 1 TO NM
150 PRINT "Density of material", I
160 INPUT R(I)
170 PRINT "Sound speed for material",I
180 INPUT V(I)
190 PRINT "Hugoniot s-coeff. for Material",I
200 INPUT S(I)
210 NEXT I
220 INPUT "Impactor velocity?",UI
230 FOR I = 1 TO NM
240 A(I) = R(I)*V(I): B(I) = R(I)*S(I)
250 NEXT I
260 U(0) = UI/2
270 FOR I = 1 TO NM-1
280 Z1 = A(I)+4*B(I)*U(I-1)+A(I+1)
290 Z2 = (A(I)+2*B(I)*U(I-1))*U(I-1)
300 Z3 = 2*(B(I)-B(I+1))
310 IF Z3 = 0 THEN 340
320 Z4 = SQR(Z1*Z1-4*Z3*Z2)
330 U(I) = (Z1-Z4)/Z3: GOTO 350
340 U(I) = 2*Z2/Z1
350 D(I) = V(I+1)+S(I+1)*U(I)
360 P(I) = R(I+1)*D(I)*U(I)
370 NEXT I
380 U(0) = U(1): D(0) = V(1)+S(1)*(UI-U(1))
390 P(0) = R(1)*D(0)*(UI-U(1))
400 PRINT
410  PRINT  "Matl1";TAB(8);  "Rho";TAB(8);"C";  TAB(8);"S":PRINT
420 FOR I = 1 TO NM
430 PRINT I, R(I), V(I), S(I)
```

```
440 NEXT I
450 PRINT
460 PRINT "Impactor velocity =";UI
470 PRINT
480 PRINT "Matl1";TAB(8);"u";TAB(8);"D";TAB(8);"P":PRINT
490 FOR I = 0 TO NM-1
500 PRINT I+1, U(I), D(I), P(I)
510 NEXT I
520 INPUT "Try another velocity? (y or n)";YN$
530 IF YN$ = "y" THEN 220
540 END
```

C.2 Direct Collision Shock Prediction

It is useful to be able to predict the shock pressure and particle velocity expected in the direct collision of a flyer and a stationary target whose Hugoniots are known. One can determine the expected data range to set up for with diagnostic equipment, etc. or predict what flyer velocity is needed to achieve a given pressure in the target. As before, we simplify nomenclature:

$$a_i \equiv \rho_{0i} c_{0i} \,, \quad b_i \equiv \rho_{0i} s_i \qquad i = 1, 2$$

Continuity of pressure at the contact surface between the flyer and target then gives

$$\rho_{01}\left[c_{01} + s_1 (u_I - u)\right](u_I - u) = \rho_{02}(c_{02} + s_2 u)u$$

Figure C.2 illustrates the analysis. Solution of the resulting quadratic for the particle velocity u_1 then gives

$$u_1 = \frac{(a_1 + 2b_1 u_I + a_2) - \sqrt{(a_1 + 2b_1 u_I + a_2)^2 - 4(b_1 - b_2)(a_1 + b_1 u_I)u_I}}{2(b_1 - b_2)}$$

When $b_1 = b_2$, the solution fails. The quadratic equation simplifies to a linear equation:

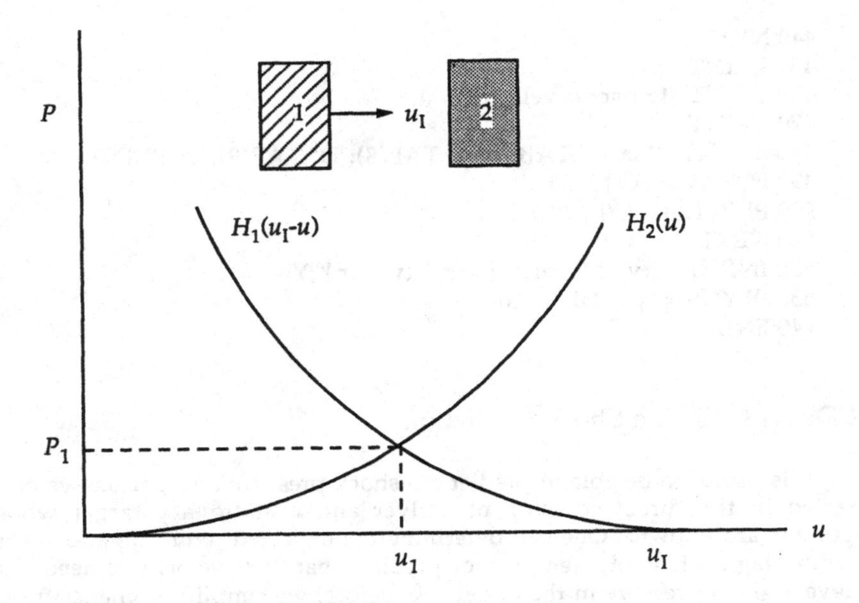

Fig. C.2. Prediction of the shocked state for a given impact velocity in a direct collision of a flyer and a stationary target when the Hugoniots for both are known.

$$(a_1 + b_1 u_I) u_I = (a_1 + 2b_1 u_I + a_2) u$$

with the solution

$$u_1 = \frac{(a_1 + b_1 u_I) u_I}{(a_1 + 2b_1 u_I + a_2)}$$

Pressure is then given by $P_1 = \rho_{02} (c_{02} + s_2 u_1) u_1$

The following computer code performs the required analysis.

```
10 REM Direct collision prediction
20 INPUT "Impactor density?"; R1
```

```
30 INPUT "Impactor sound speed?"; V1
40 INPUT "Impactor Hugoniot s-coeff.?"; S1
50 INPUT "Target density?"; R2
60 INPUT "Target sound speed?"; V2
70 INPUT "Target Hugoniot s-coeff.?"; S2
80 INPUT "Impactor velocity?"; UI
90 PRINT
100 PRINT "Matl.";TAB(8);"Rho";TAB(8);"c";TAB(8);"s":PRINT
110 PRINT "1",R1,V1,S1:PRINT "2",R2,V2,S2:PRINT
120 PRINT "Impactor velocity =", UI:PRINT
130 A1 = R1*V1: B1=R1*S1: A2 = R2*V2: B2 = R2*S2
140 Z1 = A1+2*B1*UI+A2
150 Z2 = (A1+B1*UI)*UI
160 Z3 = 2*(B1-B2): IF Z3 = 0 THEN 190
170 Z4 = SQR(Z1*Z1-2*Z3*Z2)
180 U = (Z1-Z4)/Z3: GO TO 200
190 U = Z2/Z1
200 D = V2+S2*U
210 P = R2*D*U
220 PRINT "u";TAB(8);"D";TAB(8);"P":PRINT
230 PRINT U,D,P:PRINT
240 END
```

C.3 Direct Collision Data Reduction

If the Hugoniot of a suitable flyer is known, it can be used to determine the Hugoniot of a target, one point at a time, through direct collision experiments. One measures the flyer impact velocity and the shock velocity in the target. As before, continuity of pressure at the contact surface gives

$$P = \rho_{02} D u = \rho_{01}\left[c_{01} + s_1 (u_I - u)\right](u_I - u)$$

and as before, we have

$$a_i \equiv \rho_{0i} c_{0i}\,, \quad b_i \equiv \rho_{0i} s_i \qquad i = 1, 2$$

Solution of the quadratic in u gives

$$u_1 = \frac{(a_1 + 2b_1 u_I + \rho_{02} D) - \sqrt{(a_1 + 2b_1 u_I + \rho_{02} D)^2 - 4b_1 (a_1 + b_1 u_I) u_I}}{2b_1}$$

$$P_1 = \rho_{02} D u_1$$

The following computer code performs the required analysis.

```
10 REM Direct collision data reduction
20 INPUT "Impactor density?", R1
30 INPUT "Impactor sound velocity?", V1
40 INPUT "Impactor Hugoniot s-coeff.?", S1
50 INPUT "Target density?", R2
60 INPUT "Impactor velocity?", UI
70 INPUT "Measured shock speed?", D
80 PRINT
90 PRINT "Matl1";TAB(8);"Rho";TAB(8);"c";TAB(8);"s":PRINT
100 PRINT "1",R1,V1,S1:PRINT "2",R2:PRINT
110 PRINT "Impactor velocity =";UI
120 PRINT "Measured shock speed =", D:PRINT
130 A1 = R1*V1: B1 = R1*S1
140 Z1 = A1+2*B1*UI+R2*D
150 Z2 = 4*B1*(A1+B1*UI)*UI
160 Z3 = SQR(Z1*Z1-Z2)
170 U = (Z1-Z3)/(2*B1)
180 P = R2*D*U
190 PRINT "u";TAB(8);"P":PRINT
200 PRINT U,P:PRINT
210 END
```

C.4 Impedance Match Analysis

Once the Hugoniot of a suitable material for a baseplate is known, it can be used as a standard, and the Hugoniots of other materials can be measured relative to it. A flyer, whose Hugoniot is also accurately known, impacts the baseplate material. A sample of the material whose Hugoniot is to be determined is placed on the back surface. Figure C.4 illustrates the geometry and the analysis.

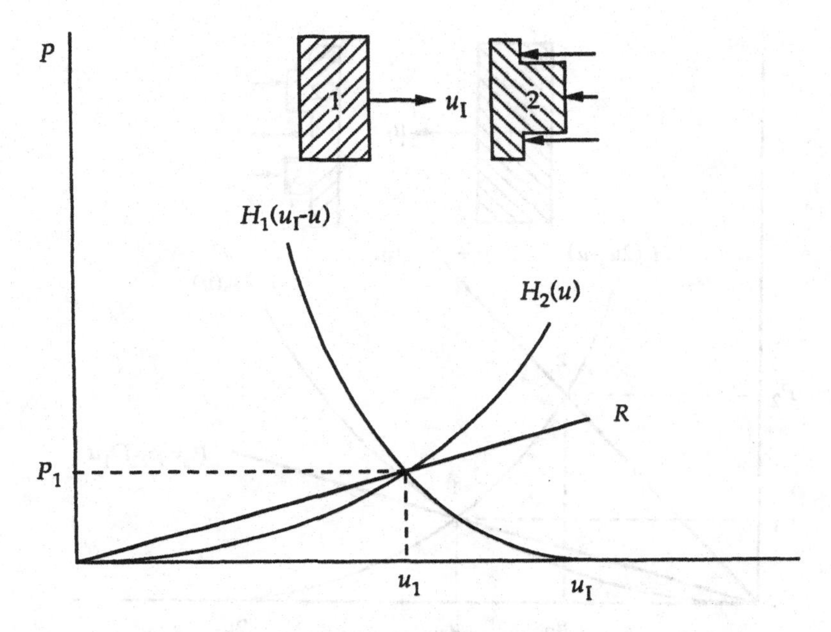

Fig. C.3. Measurement of the target Hugoniot by the direct collision method. The arrows on the target denote shorting pin detectors used to determine the shock velocity D in the target. One must measure u_I and D. It is assumed that the Hugoniot of the flyer is known. R is the Rayleigh line established by conservation of momentum. The point u_1, P_1 is on the target Hugoniot.

As before,

$$a_i \equiv \rho_{0i} c_{0i} \,, \; b_i \equiv \rho_{0i} s_i \qquad i = 1, 2$$

The shock velocities D_1 and D_2 are measured. On impact of the flyer, state 1 is established. Continuity of pressure at the contact surface gives

$$P_1 = \rho_{02} D_1 u_1 = \rho_{02} (c_{02} + s_2 u_1) u_1$$

We then have $D_1 = c_{02} + s_2 u_1$ so that we can solve for the particle velocity and pressure

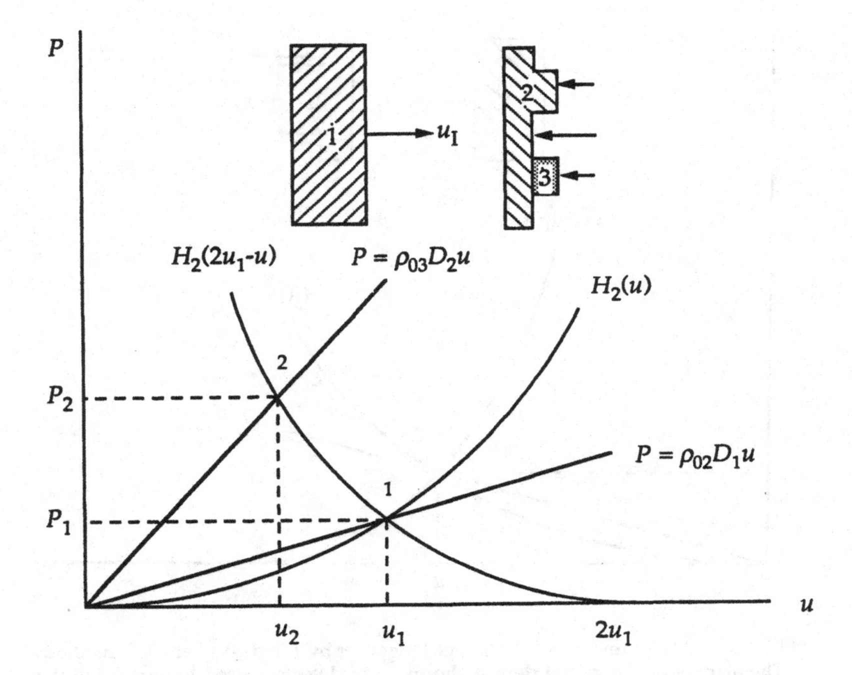

Fig. C.4. Impedance match Hugoniot measurement. The target is composed of a baseplate (material 2) whose Hugoniot is known, and a sample of the material whose Hugoniot is to be determined (material 3). The flyer Hugoniot is also known. One measures the shock velocities produced in the baseplate and in the sample. In the figure, the sample has a stiffer Hugoniot than the baseplate. Point 1 is the state resulting when the flyer strikes the baseplate. Point 2 is the state established when the shock in the baseplate arrives at the sample. It lies on the Hugoniot of the sample.

$$u_1 = \frac{D_1 - c_{02}}{s_2}, \quad P_1 = \rho_{02} D_1 u_1$$

When the shock wave reaches the sample, either reshock to a higher pressure (shown in the figure) results if the sample has a stiffer Hugoniot, or a partial release results if the baseplate Hugoniot is stiffer. Continuity of pressure at the contact surface gives

$$P_2 = \rho_{03} D_2 u_2 = \rho_{02}\left[c_{02} + s_2 (2u_1 - u_2)\right](2u_1 - u_2)$$

This can be solved for the particle velocity u_2 and the corresponding pressure P_2 can then be calculated.

$$u_2 = \frac{(a_2 + 4b_2 u_1 + \rho_{03} D_2) - \sqrt{(a_2 + 4b_2 u_1 + \rho_{03} D_2)^2 - 8b_2 (a_2 + 2b_2 u_1) u_1}}{2b_2}$$

$$P_2 = \rho_{03} D_2 u_2$$

Normally, the baseplate is chosen to have a softer Hugoniot. If it is stiffer than that of the sample, release along the baseplate isentrope from state 1 results. This can be approximated by the reflected Hugoniot in the figure, but the approximation affects the accuracy of the results for the Hugoniot of the sample.

The following computer code performs the required analysis.

```
10 REM Impedance match analysis
20 INPUT "Baseplate density?", R2
30 INPUT "Baseplate sound speed?", V2
40 INPUT "Baseplate Hugoniot s-coeff.?", S2
50 INPUT "Target density?", R3
60 INPUT "Shock speed in baseplate?", D1
70 INPUT "Shock speed in target?", D2
80 PRINT
90 PRINT "Matl1";TAB(8);"Rho";TAB(8);"c";TAB(8);"s":PRINT
100 PRINT "2",R2,V2,S2:PRINT "3",R3:PRINT
110 PRINT "Measured shock speeds":PRINT
120 PRINT "Baseplate";TAB(8);"Target":PRINT
130 PRINT D1,D2: PRINT
140 U1 = (D1-V2)/S2:P1 = R2*D1*U1
150 A2 = R2*V2: B2 = R2*S2
160 Z1 = A2+4*B2*U1+R3*D2
170 Z2 = 8*B2*(A2+2*B2*U1)*U1
180 Z3 = SQR(Z1*Z1-Z2)
190 U2 = (Z1-Z3)/(2*B2)
```

```
200 P2 = R3*D2*U2
210 PRINT TAB(8);"Baseplate";TAB(8);"Target":PRINT
220 PRINT "u", U1, U2
230 PRINT "P",P1,P2:PRINT
240 END
```

C.5 Reshock Analysis

One can measure higher pressures on the Hugoniot of a target by backing it with a material with a stiffer Hugoniot, so that a reflected shock results in the target when the initial shock reaches the backplate. It has been found experimentally (there has been no theoretical justification), that the reshock Hugoniot emanating from the initial shocked state on the target Hugoniot is a reflection of the principal Hugoniot in the particle velocity resulting from the initial shock. One drawback of the technique is that the accuracy of Hugoniot points determined by this method suffers from the compound nature of the process. Errors involved in determining the initial shocked state are multiplied in the second step. Nevertheless, when there is no other way to reach the desired pressure, this technique is useful. Both the flyer and backplate Hugoniots must be accurately known. One measures the flyer velocity and the shock velocity in the backplate. Figure C.5 illustrates the geometry and the analysis. On initial impact, continuity of pressure at the contact surface gives

$$\rho_{02}(c_{02}+s_2 u_1)u_1 = \rho_{01}\left[c_{01}+s_1(u_f-u)\right](u_f-u)$$

Solution of the quadratic for u_1 gives

$$u_1 = \frac{(a_1+2b_1 u_f+a_2)-\sqrt{(a_1+2b_1 u_f+a_2)^2-4(b_1-b_2)(a_1+b_1 u_f)u_f}}{2(b_1-b_2)}$$

$$P_1 = \rho_{02}(c_{02}+s_2 u_1)u_1$$

For $b_1 = b_2$ this solution fails. We then have

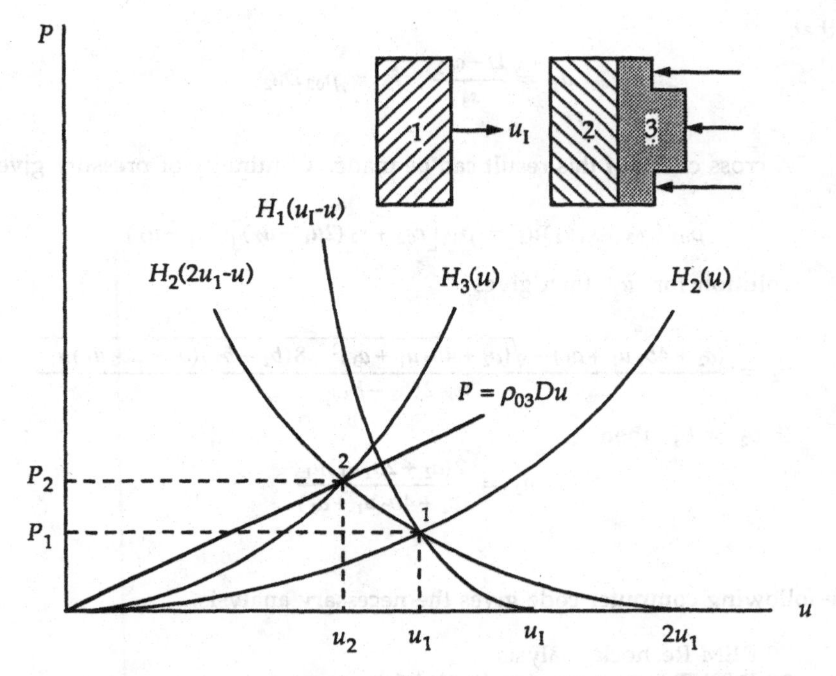

Fig. C.5. Reshock analysis. Material 2 is the sample. When the initial shock arrives at the backplate 3, which has higher shock impedance and a well known Hugoniot, a reflected shock is sent back into the sample. Point 1 is the initial shocked state. Point 2 is the reshocked state. It has been found by experiment that the continuation of the Hugoniot $H_2(u)$ beyond point 1 is a mirror image of $H_2(2u_1\text{-}u)$ in point 1.

$$(a_1 + 2b_1 u_I + a_2)u_1 = (a_1 + b_1 u_I)u_I$$

and then

$$u_1 = \frac{(a_1 + b_1 u_I)u_I}{(a_1 + 2b_1 u_I + a_2)}$$

Now on reshock, continuity of pressure gives

$$\rho_{03} D u_2 = \rho_{03}(c_{03} + s_3 u_2)u_2$$

so that

$$u_2 = \frac{D - c_{03}}{s_3}, \quad P_2 = \rho_{03} D u_2$$

A cross check of this result can be made. Continuity of pressure gives

$$\rho_{03}(c_{03} + s_3 u_2) u_2 = \rho_{02}\left[c_{02} + s_2(2u_1 - u_2)\right](2u_1 - u_2)$$

Solution for u_2 then gives

$$u_2 = \frac{(a_2 + 4b_2 u_1 + a_3) - \sqrt{(a_2 + 4b_2 u_1 + a_3)^2 - 8(b_2 - b_3)(a_2 + 2b_2 u_1) u_1}}{2(b_2 - b_3)}$$

If $b_2 = b_3$ then

$$u_2 = \frac{2(a_2 + 2b_2 u_1) u_1}{(a_2 + 4b_2 u_1 + a_3)}$$

The following computer code gives the necessary analysis.

```
10 REM Reshock analysis
20 INPUT "Impactor density?", R1
30 INPUT "Impactor sound speed?", V1
40 INPUT "Impactor Hugoniot s-coeff.?", S1
50 INPUT "Target density?", R2
60 INPUT "Target sound speed?", V2
70 INPUT "Target Hugoniot s-coeff.?", S2
80 INPUT "Anvil density?", R3
90 INPUT "Anvil sound speed?", V3
100 INPUT "Anvil Hugoniot s-coeff.?", S3
110 INPUT "Impactor velocity?", UI
120 INPUT "Shock speed in anvil?", D
130 PRINT
140 PRINT "Matl1";TAB(8);"Rho";TAB(8);"c";TAB(8);"s":PRINT
150 PRINT "1",R1,V1,S1:PRINT "2",R2,V2,S2:PRINT "3",R3,V3,S3
160 PRINT
170 PRINT "Impactor velocity =", UI
180 PRINT "Measured shock speed =", D:PRINT
190 A1=R1*V1:B1=R1*S1:A2=R2*V2:B2=R2*S2:A3=R3*V3:B3=R3*S3
```

```
200 Z1 = A1+2*B1*UI+A2
210 Z2 = (A1+B1*UI)*UI
220 Z3 = 2*(B1-B2): IF Z3 = 0 THEN 250
230 Z4 = SQR(Z1*Z1-2*Z3*Z2)
240 U1 = (Z1-Z4)/Z3: GOTO 260
250 U1 = Z2/Z1
260 P1 = R2*(V2+S2*U1)*U1
270 U2 = (D-V3)/S3:P2 = R3*D*U2
280 Z5 = A2+4*B2*U1+A3
290 Z6 = (A2+2*B2*U1)*U1
300 Z7 = 2*(B2-B3): IF Z7 = 0 THEN 330
310 Z8 = SQR(Z5*Z5-4*Z7*Z6)
320 UC = (Z5-Z8)/Z7: GOTO 340
330 UC = 2*Z6/Z5
340 PRINT
350 PRINT "u1";TAB(8);"P1";TAB(8);"u2";TAB(8);"P2":PRINT
360 PRINT U1,P1,U2,P2:PRINT
370 PRINT "Check u2 =", UC:PRINT
380 END
```

C.6 Calculation of Edge Rarefaction Effects on an Embedded Detector.

Embedded detectors are sometimes used for velocity gauges and for other purposes. All impactors are of finite extent. As a result, edge rarefactions from the side of the impactor eventually merge behind the shock front to attenuate it. It is important to be able to determine when an edge rarefaction arrives at the detector, causing lateral motion. Recorded data up to that time correspond to linear motion of the gauge in the direction of the shock front, simplifying analysis. Figure C.6 illustrates the calculation. It is assumed in the calculation that the impactor is thick enough that the overtaking rarefaction from the rear of the impactor does not reach the detector first. Point 1 is the initial position of the detector. Point 2 is the position of the detector when the edge rarefaction catches it. We will calculate an approximate result by approximating the sound speed behind the shock by the shock speed ($c \approx D$) , which is often a very good approximation. The shock is assumed to arrive at the detector at time t_1 . The rarefaction from the edge of the impactor arrives at time t_2 . The period of one-dimensional motion at the detector is thus $t_2 - t_1$.

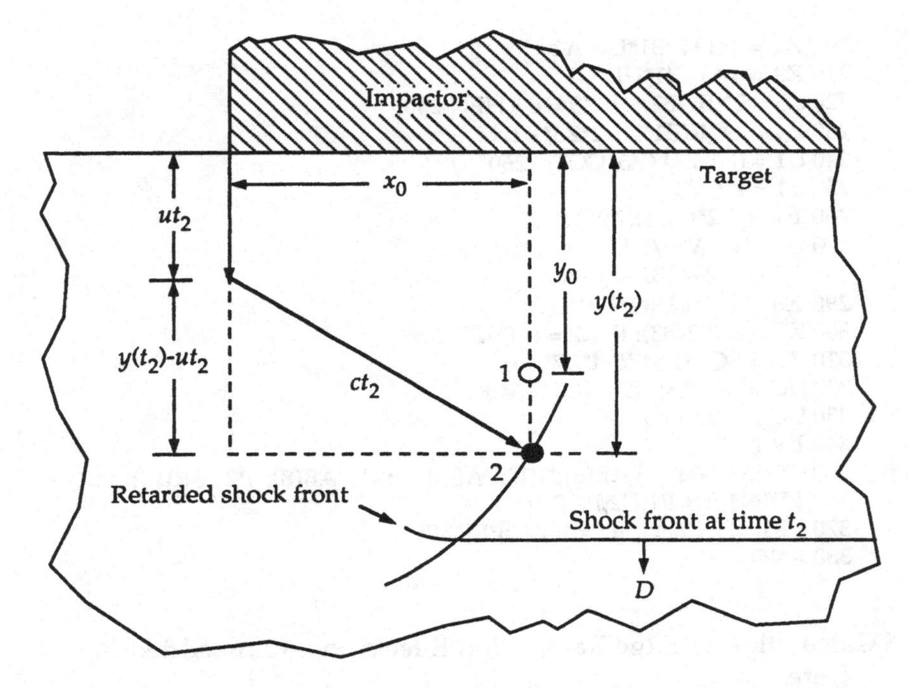

Fig. C.6. Calculation of the duration of one dimensional motion at an embedded detector in a target. Point 1 is the initial position of the detector. Point 2 is the position of the detector after moving with the particle velocity behind the shock. At time t_2 a release wave from the edge of the impactor reaches the detector.

From the figure we have

$$t_1 = \frac{y_0}{D}$$

and using Pythagorean theorem in the lower left triangle in the figure, we have $c^2 t_2^2 = x_0^2 + \left[y(t_2) - ut_2\right]^2$. Also, we have $y(t_2) = y_0 + u(t_2 - t_1)$, so that $y(t_2) - ut_2 = y_0 - ut_1$. Substituting for the time t_1 we thus have $y(t_2) - ut_2 = y_0(1 - u/D)$. We thus have

$$t_2 = \frac{\sqrt{x_0^2 + y_0^2\left(1-\frac{u}{D}\right)^2}}{c}$$

Using our approximation $c \approx D$ for the sound speed behind the shock this becomes

$$t_2 \cong \frac{\sqrt{x_0^2 + y_0^2\left(1-\frac{u}{D}\right)^2}}{D}$$

The following computer code performs the required calculations.

```
10 REM Edge rarefaction code
20 INPUT "Impactor density?", R1
30 INPUT "Impactor sound speed?", V1
40 INPUT "Impactor Hugoniot s-coeff.?", S1
50 INPUT "Target density?", R2
60 INPUT " Target sound speed?", V2
70 INPUT "Target Hugoniot s-coeff.?", S2
80 INPUT "Impactor velocity?", UI
90 PRINT
100 PRINT "Matl1";TAB(8);"Rho";TAB(8);"c";TAB(8);"S":PRINT
110 PRINT "1",R1,V1,S1:PRINT "2",R2,V2,S2:PRINT
120 PRINT "Impactor velocity =",UI:PRINT
130 A1 = R1*V1: B1 = R1*S1: A2 = R2*V2: B2 = R2*S2
140 Z1 = A1+2*B1*UI+A2
150 Z2 = (A1+B1*UI)*UI
160 Z3 = 2*(B1-B2): IF Z3 = 0 THEN 190
170 Z4 = SQR(Z1*Z1-2*Z3*Z2)
180 U = (Z1-Z4)/Z3: GOTO 200
190 U = Z2/Z1
200 D = V2+S2*U: P = R2*D*U
210 PRINT "u";TAB(8);"D";TAB(8);"P":PRINT
220 PRINT U,D,P:PRINT
230 INPUT "Depth of Gauge?", Y
240 INPUT "Distance from the edge?", X
250 PRINT
260 PRINT "Depth";TAB(8);"Distance from edge":PRINT
270 PRINT Y,X:PRINT
280 Z5 = SQR(X*X+Y*Y*(1-U/D)*(1-U/D))
290 T1 = Y/D
```

```
300 T2 = Z5/D
310 PRINT:PRINT
320 PRINT "t1";TAB(8);"t2":PRINT
330 PRINT T1,T2:PRINT
340 INPUT "Try another geometry?" (y or n)",YN$
350 IF YN$ = "y" then 230
360 END
```

Appendix D. Selected Shock and Thermodynamic Properties

The following table of approximate properties was originally distributed at the 1989 Topical Conference on Shock Waves in Condensed Matter, sponsored by the American Physical Society.

Table D.1 Selected Shock and Thermodynamic Properties of Materials

Material	$\rho(g/cm^3)$	$c_0(cm/\mu s)$	s	$c_p(J/g \bullet K)$	γ_G
Ag	10.49	0.323	1.60	0.24	2.5
Au	10.24	0.306	1.57	0.13	3.1
Be	1.85	0.800	1.12	0.18	1.2
Bi	9.84	0.183	1.47	0.12	1.1
Ca	1.55	0.360	0.95	0.66	1.1
Cr	7.12	0.517	1.47	0.45	1.5
Cs	1.83	0.105	1.04	0.24	1.5
Cu	8.93	0.394	1.49	0.40	2.0
Fe[a]	7.85	0.357	1.92	0.45	1.8
Hg	13.54	0.149	2.05	0.14	3.0
K	0.86	0.197	1.18	0.76	1.4
Li	0.53	0.465	1.13	3.41	0.9
Mg	1.74	0.449	1.24	1.02	1.6
Mo	10.21	0.512	1.23	0.25	1.7
Na	0.97	0.258	1.24	1.23	1.3
Nb	8.59	0.444	1.21	0.27	1.7
Ni	8.87	0.460	1.44	0.44	2.0
Pb	11.35	0.205	1.46	0.13	2.8
Pd	11.99	0.395	1.59	0.24	2.5
Pt	21.42	0.360	1.54	0.13	2.9
Rb	1.53	0.113	1.27	0.36	1.9
Sn	7.29	0.261	1.49	0.22	2.3
Ta	16.65	0.341	1.20	0.14	1.8
U	18.95	0.249	2.20	0.12	2.1
W	19.22	0.403	1.24	0.13	1.8
Zn	7.14	0.301	1.58	0.39	2.1

Table D.1 Selected Shock and Thermodynamic Properties of Materials (continued)

material	ρ(g/cm^3)	c_0(cm/μs)	s	c_p(J/g•K)	γ_G
KCl[a]	1.99	0.215	1.54	0.68	1.3
LiF	2.64	0.515	1.35	1.50	2.0
NaCl[b]	2.16	0.353	1.34	0.87	1.6
Al - 2024	2.79	0.533	1.34	0.89	2.0
Al - 6061	2.70	0.535	1.34	0.89	2.0
SS - 304	7.90	0.457	1.49	0.44	2.2
Brass	8.45	0.373	1.43	0.38	2.0
Water	1.00	0.165	1.92	4.19	0.1
Teflon	2.15	0.184	1.71	1.02	0.6
PMMA[c]	1.19	0.260	1.52	1.20	1.0
PE[d]	0.92	0.290	1.48	2.30	1.6
PS[e]	1.04	0.275	1.32	1.20	1.2

a above phase transition, b below phase transition, c polymethylmethacrylate, d polyethylene, e polystyrene

Appendix E: Table of Symbols and Units

The following tables list the units used in this book, and the principle usage of the variable symbols. On occasion, a symbol may be temporarily defined to mean something else (eg; μ, λ = Lamé constants, $\beta = 1/kT$). In such cases, the definition is given locally where it is used. Primes on a variable indicate either a total derivative or in some cases, a new value.

Units:

m = meter
g = gram
s = second
J = Joule
Bar = 10^6 dyne/cm^2 = 0.1 MPa
Pa = Pascal = 10 dyne/cm^2

Modifiers:

μ = micro, m = milli, c = centi, k = kilo, M = mega, G = Giga

Symbols:

β = isobaric bulk thermal expansion coefficient (also $1/kT$)
B_{0s} = adiabatic bulk modulus at normal conditions
B_T = isothermal bulk modulus
c = sound speed
c_p = specific heat (per unit mass) at constant pressure
c_v = specific heat (per unit mass) at constant volume
c_0 = intercept for Hugoniot fits
D = Shock velocity
Δ = an increment
δ = a small increment, or an inexact differential
δ_{ij} = 1 for i = j, zero otherwise

E	=	internal energy/mass
E_{coh}	=	cohesive energy
ε	=	dimensionless internal energy
e_{ij}	=	elasticity tensor
EOS	=	equation of state
F	=	Helmholtz Free energy
fcc	=	face centered cubic
γ	=	specific heat ratio c_p/c_v
γ_G	=	Grüneisen coefficient
γ_0	=	Grüneisen coefficient at normal conditions
h	=	specific enthalpy
H	=	often used to denote the Hugoniot as a function
K	=	degrees Kelvin
κ_T	=	isothermal compressibility
L	=	Latent heat of phase change
λ	=	One of the Lamé constant (also fraction of a phase change)
μ	=	compressibility factor $\rho/\rho_0 - 1$ (also a Lamé constant)
P	=	pressure
P_c	=	Pressure on the cold (0 K) curve or P at the critical point)
q	=	quantity of heat
R	=	Universal gas constant
ρ	=	mass density $1/V$
ρ_{oR}	=	reference density (not necessarily normal density)
S	=	specific entropy
S	=	Shock strength (pressure ratio)
s	=	slope coefficient for Hugoniot fits
σ	=	soft-sphere diameter (also Stephan-Boltzmann constant)
t	=	time
τ_{ij}	=	stress tensor
T_c	=	Temperature at the critical point
u	=	particle velocity
u_I	=	impactor velocity
V	=	volume/mass
V_c	=	Specific volume at the critical point
Y	=	Yield stress
Z_c	=	compressibility ratio $P_c V_c / T_c$ at the critical point

Subscripts:

c	=	value on the cold (0 K) curve or at the critical point
EOS	=	denotes values calculated from the equation of state
H	=	denotes values on the Hugoniot
I	=	value for an impactor
ph	=	value on the phase boundary
0	=	value at initial or normal conditions

Index

The following index was constructed using the finder in the word processor software WORD 4.0, which highlights items found, but does not explicitly indicate the exact page where the item is located. As a result, the indicated pages may at times be a page away from where the item is located in the manuscript. The scan did not include the foreword or the table of contents.